Global Foodscapes

What we eat – as well as how it is produced, processed, moved, sold, and used by our bodies – seems to matter like never before. *Global Foodscapes* takes on this topicality and asks readers to think about how we are all involved in the making of an odd and, in many ways, troubling and contested food economy. It explores how food is conceived, traded, grown, reared, processed, sold, and consumed; investigates what goes wrong along the way; and assesses what diverse people around the world are doing to fix these faults.

The text uses a carefully crafted framework that explores the interaction of five forms of oppression and five means of resistance as they are worked out over five stages in the food economy. It draws on case studies from around the world that illuminate key issues about food in today's world; examines how oppression affects diverse people caught up in the food economy; and highlights how individuals, groups, and institutions such as governments, but also firms, are trying to improve how we interact with the food system.

Global Foodscapes is a highly accessible and useful text for undergraduate students interested in the global food economy. The global range of case studies, examples, and reference points, as well as its original framework, allows the text to speak to diverse audiences and generate debate about whether anything – and if so, what – needs to be done about the food system we depend upon so heavily. Additional materials such as suggested readings and discussion points help students consider the issues at hand and conduct initial and more detailed research on today's food economy.

Alistair Fraser is a Lecturer in the Department of Geography, Maynooth University, Ireland. His research is diverse in nature – cutting across political, economic, and cultural geography – and includes journal articles on land reform in South Africa, scale in political practice, and music.

Global Foodscapes

Oppression and resistance in the life of food

Alistair Fraser

LONDON AND NEW YORK

First published 2017
by Routledge
2 Park Square, Milton Park, Abingdon, Oxon OX14 4RN

and by Routledge
711 Third Avenue, New York, NY 10017

Routledge is an imprint of the Taylor & Francis Group, an informa business

British Library Cataloguing in Publication Data
A catalogue record for this book is available from the British Library

Library of Congress Cataloging in Publication Data
Names: Fraser, Alistair, 1982- author.
Title: Global foodscapes : oppression and resistance in the life of food / Alistair Fraser.
Description: Abingdon, Oxon ; New York, NY : Routledge is an imprint of the Taylor & Francis Group, an Informa Business, [2017] | Includes bibliographical references and index.
Identifiers: LCCN 2016013319| ISBN 9781138192478 (hardback : alk. paper) | ISBN 9781138192485 (pbk. : alk. paper) | ISBN 9781315639710 (ebook)
Subjects: LCSH: Food industry and trade--Social aspects. | Food supply--Social aspects. | Food supply--Economic aspects.
Classification: LCC HD9000.5 .F69 2017 | DDC 338.1/9--dc23
LC record available at https://lccn.loc.gov/2016013319

ISBN: 978-1-138-19247-8 (hbk)
ISBN: 978-1-138-19248-5 (pbk)
ISBN: 978-1-315-63971-0 (ebk)

Typeset in Sabon
by Saxon Graphics Ltd, Derby

Printed and bound in Great Britain by
TJ International Ltd, Padstow, Cornwall

For Veronica, Emiliano, and Fabio.

Contents

Figures

Tables

Acknowledgements

My parents and then my siblings Shona, Shari, and Gordon have given me extraordinary and enduring support over the years. Now my own family – Veronica and our boys Emiliano and Fabio – do the same, in new ways. I am fortunate to have such wonderful people so close to me. Veronica, who has encouraged me to write this book and helped keep me in a positive frame of mind throughout, has also given me so much love, for which I want to thank her here. I also want to thank everyone in Veronica's family, and especially her mother, Yolanda, for all she has done to help us over the last three years in particular. Also due warm thanks here are my friends Delphine Ancien, Jean Clarke, Will Dowling, Sandra Hayes, Tim Jolly, and Roy, Sean, Breffni, and Stuart. En Mexico, quiero dicir 'muchas gracias' a Juanita.

Within the academic world three people in particular have helped me over the years. In Scotland in the late 1990s and now in Ireland, Mark Boyle has always offered me wise advice and lent me a hand when I've needed it. At Ohio State, Kevin Cox and Nancy Ettlinger have consistently read my work with a level of enthusiasm and care that all would-be writers deserve. Not enough of what they have been able to teach me appears in the following pages. I also thank Álvaro López López, María del Carmen Juárez Gutiérrez, and Lilia Susana Padilla y Sotelo in the Instituto de Geografía, Universidad Nacional Autónoma de México, for giving me space in which to work – and time to practise my wonky Spanish skills – during my sabbatical leave.

In Maynooth I am fortunate to have some kind and stimulating colleagues. I must particularly thank Jan Rigby, Cian O'Callaghan, Sinéad Kelly, Chris Van Egeraat, Gerry Kearns, and Karen Till. Paddy Duffy helped me in many ways when I arrived in Maynooth; his intellectual curiosity and humility make him the ideal role model for a rather brash person like me. Also helpful over the years have been the students I have taught in Maynooth. Special thanks must go to Annette Egan, Colin McCarthy, and Colm Maloney. More generally, all my students in GY340 and GY333 deserve thanks for putting up with me as I tried to get my ideas worked out.

In producing the book, I thank Maynooth University for financial support for some of the publishing costs (and for continuing to value the concept of sabbatical leave). I would also like to thank Andrew Mould and Egle Zigaite at Routledge. Thoughtful and generous comments from anonymous referees have encouraged me to make the text stronger. Kevin Cox and Nancy Ettlinger carefully read earlier drafts of the chapters here. Gerry Kearns gave me advice about the publishing process. I would like to thank the United Nations and the World Health Organization for granting me permission to use some extracts from their published work. Finally, thank you Doc Scott and Loxy.

Introduction

Introduction

My aim in this book is to present a diverse but connected set of materials regarding what I refer to as 'foodscapes'. What I mean by focusing on this idea of foodscapes is that, to eat – that is, for food to get onto our tables, into restaurants, canteens, supermarkets, or convenience stores – a vast number of spaces and places must be produced and connected, not only with one another but also with a wide range of people all across the world. Foodscapes is the word I use to refer to these spaces and places (for some other examples of how scholars have used the term 'foodscape', see Goodman *et al.*, 2010; Brembeck *et al.*, 2013). Foodscapes, to me, can be small: your lunch box, my desk, the table on which we have dinner. And they can be as large as a restaurant, farm, or biscuit factory. Some are, as it were, stretched out, sprawling, connecting action in multiple sites – perhaps the foodscape of a large transnational firm comes to mind. Others are bunched-up, constituted by social actors using goods and services purchased from near by to make new goods – food crops, granola, bread – that are sold locally.

It is true that all of these sorts of foodscapes are always local to someone – the strip mall near your house where Burger King or Domino's Pizza are located; the downtown, maybe gentrified, neighbourhood where you can find nice restaurants, bars, and maybe some cool food trucks; or the farm shop you pass on the way to work or to a national park. These sorts of foodscapes will usually need workers drawn from the local area, if for no other reason than because the food on offer has to be unpacked, prepared, heated, and served. In a sense, then, I could have titled this book 'Local Foodscapes'; I even could have dropped any reference to some particular geographical scale. However, I want to propose that we need to see foodscapes as always in some way *global*. Why?

On the one hand, so much of the food and drink we consume these days is brought from far away. Just today, here in Ireland where I am writing this, I ate some grapefruit in a fruit salad at breakfast, as well as some French cheese, a bag of peanuts I bought in Mexico, and then three cups of coffee using beans grown on coffee plants, and picked by people living in Veracruz, Mexico. In Mexico, where I have spent a fair bit of time living in the last two years, I sometimes eat Uruguayan meat, add oats grown in the US to the bread I make, and when I feel like a treat I have some matured Gouda cheese from Holland. The food I eat has this global character: drawn from diverse places around the world, shipped to my local supermarket or convenience store, and effectively making distant people and their labour involved in my eating. Admittedly, I lead a privileged life and can afford such luxuries as matured Gouda, which is expensive. But even some of the poorest people in the world eat wheat or rice brought into their countries from the other side of the world, or buy processed foodstuffs based on US or European dairy products. Moreover, many of the world's poorest people live close to plantations or roads or ports where crops are grown and packed for export to rich consumers around the world: produce passes them by, on its way to a distant market, leaving behind many who would like, but cannot afford, to eat the avocados or mangoes destined for European or Canadian consumers. Many of the foodscapes we see around us – at home, work, or when we eat out – are similarly constituted. They have this international dimension, this 'global*ness*' to them; hence I think it is necessary for us to dwell upon the way foodscapes today are *global* foodscapes.

Beyond this point about the vast distances and diverse movements of foodstuffs today, there is something slightly more profound – and complex – about contemporary foodscapes that make it important for us to view them as *global*. Today, around 780 million people are undernourished in so-called 'developing' countries; and there are another 14.7 million undernourished people in the 'developed' world (Food and Agriculture Organization, 2015: 44). To be undernourished means that, for at least one year, an individual has been unable to acquire enough food to meet their dietary needs. This situation is hard to imagine for someone like me – and possibly you – who has never had to go hungry. Yet, my eating (and the spaces and places that must be produced and connected with one another to get food onto my table) has to be seen as interlinked with hunger. It might be tempting to divorce these aspects of contemporary life, and I suspect some of us *do* avoid thinking about it. But hunger on the other side of the street, down the road, in the city where I live, as well as on the other side of the world, is produced in large part by the same general social processes as those that allow me to have enough money to shop at the supermarket and eat out in a restaurant. Today, indeed, these general social processes unfolding in all of the different places around the world are so bound up with one another – so interwoven, folded into each other, so tangled-up – that they constitute a global political and economic society. The diverse and multiple movements of commodities of all shapes and sizes across the world, the new map of connections created by migration flows, or that crumpled way in which the world has been shrunk by today's telecommunications technologies: these developments make our lives and all of the various foodscapes we can see and touch and feel around us *global*.

In *Global Foodscapes*, then, my aim is to pay attention to these landscapes of food. I want to critically explore a range of problematic, as well as some promising, features emerging from the complex array of social processes that combine to get food onto the market and into our stomachs. Before beginning, however, it is necessary to set the scene and so, in this introductory chapter, I want to discuss three issues. The first of these is about 'capitalism' and why it is so fundamental to what we need to cover here. Specifically, I think we need to grasp how capitalist firms – or we might call them 'businesses' and, when appropriate, 'corporations' – shape foodscapes via their pursuit of profit. These firms have become critical producers, movers, and sellers of the commodities needed to produce food. And they buy, trade, move, process and sell food. They are, as such, central to the daily challenge of getting sufficient food into the towns and cities where so many of us, divorced from the land and relying on a steady supply of food coming to us, now live. Capitalism matters. The second issue is about this idea of mine that we can detect – and need to think about – a range of problematic and promising features emerging from the global foodscape. To be more specific, I discuss how oppression and resistance are two key outcomes of contemporary society, capitalist or not, and that global foodscapes today are constituted in part by the interplay between them. A focus on oppression and resistance highlights what is going wrong in these landscapes of food, while also shining a light on what some actors are trying to do about it. My point is that spaces and places in general – that is, whether we are interested in foodscapes or some other spatial concept – are always products of messy power plays: some actors can get what they want; others lose out. I think shining a light on power relations and how they get worked out in diverse ways can help us understand how we relate to foodscapes today. The third and final issue is about how I explore the 'life of food'. I see this unfolding in five stages: from a world upstream of the farm all the way downstream to the moment when we eat and our bodies begin to digest and use up food in our stomachs. In between, there is food production on the land, food processing in bakeries or factories, and then the selling of food in diverse and sometimes rather strange spaces. All along the way, governments, capitalist firms, farmers, people like you and me, as well as a wide range of affected groups, work together and maybe even against each other to shape the foodscapes we experience around us. The net result, I think, has been the emergence of intriguing, complex, and challenging global foodscapes.

Capitalism matters

Capitalism matters enormously when it comes to the way foodscapes are produced. Firms, businesses, corporations – capitalists – shape how food is produced, moved, and sold; and they see ways to profit from this. They make goods or provide services that farmers will use and food products using the crops coming off the farm; sell food to us wherever we happen to be; and take part in the actual process of digesting and metabolizing food. In a sense, capitalists want to get into our stomachs. That these firms have become so central to the way we eat is undeniable. In turn, this means we need to develop an account of how this has become possible. What is it about contemporary society that enables firms like these to become so powerful?

Globally, contemporary societies are diverse. There are parliamentary democracies, monarchies, and Islamic Republics; liberal societies with progressive laws, as well as conservative societies with repressive ones; agrarian and industrial societies; and so on. Diverse, then. Yet, in all of these societies there is a type of enterprise that we can quite justifiably refer to as 'capitalist' because it seeks to profit from production carried out by its employees. In such a firm, workers sign up to earn a wage by working a set number of hours. They work with machinery or on a computer, sometimes alone but usually with others. Some might have a good laugh while they work; others will detest it. But in the process of working those hours, some profit should emerge, albeit if the owner or manager has done things correctly and if the end product – whatever it may be – can be sold at a price that covers what was paid out in the first place for raw materials, tools, and labour, and leaves some left over for the owner(s). Because there is a surplus produced during the working day – a profit – which the employer keeps (and can use to re-invest, distribute to shareholders, or just spend), it is important to see the exchange of wages for labour as exploitative. Workers spend longer working than they need to, were the employer only to cover their costs of making it back to work the next time. They are exploited in a capitalist firm.

The historical development of this sort of capitalist enterprise has taken place over the last three centuries. With origins in England and elsewhere in Western Europe, the capitalist firm has spread throughout the world. These sorts of enterprises compete with others over cost or quality, and in the process they have developed technological innovations that speed up production, make systems more efficient, and enable the creation of entirely new products and services. In the wake of the capitalist firm's geographical expansion have emerged societies that are, to a very large extent, subordinate to the 'logic' of the profit-making firm. In saying 'logic' I do not want to mystify things; this is not sorcery we are talking about. Rather, the point is that capitalist enterprises must continue making profits. Their owners can see future profits, almost stacked up in front of them, within reach; the task is to accumulate them. And so, if there are obstacles in the way, capitalists are always intent on getting them removed. What might these obstacles be? For one thing, consider that all firms always need labour. They need people who are looking to earn a wage; people who *must* earn a wage if they are to survive. However, a supply of labour is not always available. An obstacle. Likewise, because capitalist firms need a physical infrastructure to support production or the distribution of products, other obstacles might be encountered such as an unsteady supply of electricity or an unreliable road network. Alternatively, some firms might want a legal and regulatory environment that suits them: laws that protect their property, or rules about how the larger firms must deal with smaller ones. And given the importance of capitalist firms – the subordination of societies to the logic of capitalism – we should not be surprised to hear calls for immigration reform, infrastructural investment, or legal changes from the associations or institutions run by business. These are just some of the ways capitalist firms can put pressure on the societies in which they operate to change and become more amenable to their needs. Wherever they are based, the owners of capitalist firms are prone to call on governments and the rest of society to ensure their needs are met. Although contemporary societies throughout the world are diverse in many ways, they are all liable to become subordinated to the logic of the capitalist firm – the need to keep production going, to draw goods and services in and make new commodities; to keep making profits.

Thus, when we look at how the commodities we consume actually get to us – and in saying this, I think it is useful to think about food *as well as* non-food commodities such as our smartphones, household furniture, or the clothes we wear – what we find is a vast array of capitalist firms working in locations all over the world, pulling crops and minerals out of the ground, shipping them to market or to a factory, sometimes on the other side of the world, and creating entirely new goods that we buy in the mall, online, or wherever. Alongside them are a host of other firms producing services, such as the software and apps on our phones or laptops, or services that other firms require to stay afloat, manage logistics, and stay profitable. In combination, these firms are producing and reproducing a capitalist world, employing billions of people, getting goods to market, and making money for their owners. And because they employ people and because they make money, they are in a prime position to continue making demands on governments to generate the right sorts of conditions in which they can continue to exist and grow. When we think about the position and the centrality of capitalist firms in our world today, we should imagine a formidable and diverse complex of players that are intent on seeing the reproduction of conditions in which they continue to accumulate profit.

From the proletarian food question to the corporate food regime

Vast numbers of people living in capitalist societies have been proletarianized, i.e. they or their ancestors have been divorced from the land and must now earn a wage to survive. They are workers, even if some are well-paid, earn a salary rather than just a wage, and identify as 'middle' rather than 'working' class. Precisely because they are divorced from the land and cannot produce food to last a whole year, they are now dependent on food coming to them.

And because a reasonable but varying chunk of their earnings must be used to purchase food, they are structurally dependent on a food system that can generate food at a price they can afford. Likewise, and critically, their employers – capitalists, as individuals as well as a class – rely on the availability of food with a relatively low price because wage levels and food prices are intimately connected: if there are consistent food price increases, workers are likely to demand higher wages, or take to the streets and call on the government to take action, and even for it to be overthrown, as happened in Egypt and Syria in 2011.

The upshot of all this is what I think we can refer to as the 'proletarian food question'. This question is about whether a consistent supply of food can reach workers at a cost that is sufficiently low enough that it will not generate demand for wage increases. I want to suggest that governments and capitalist firms have sought ways of combining their efforts with agricultural sectors to answer this question day after day, month after month, year after year. In effect, the proletarian food question has given an impetus to the development of a particular view of food production: that the state and agriculture and business should work together to ensure there is enough output that food can always reach the city; and that this food should be produced at a relatively low cost so as to avoid the need to pay workers higher wages.

In the area of food production, the result is what I call 'northern agriculture' ('northern' because of its geographical origins in 'global north' countries such as the US, Japan, and those in Western Europe). As we will see, this is an input-dependent, quasi-industrial way of farming the land. It needs inputs such as fertilizer, agricultural chemicals, machinery, and waged labour; and it tends to rely heavily on wide-ranging government subsidies. 'Upstream' of agriculture is a vast array of capitalist firms producing inputs and offering various services needed to keep northern agriculture going. 'Downstream' are processors and retailers, many of them operating on an enormous scale, which means they can exert pressure on farmers to reduce prices. Viewed together, we can detect a consistent and widespread effort to reduce the relative costs of food production and distribution, even at the expense of the environment or the people who work in these diverse foodscapes.

As I will discuss, therefore, capitalist firms upstream or downstream of agriculture play an absolutely crucial role in answering the proletarian food question. But more than just playing a role, I think we need to imagine the food system as increasingly oriented around and working *for* these firms. Indeed, drawing on the work of Philip McMichael (e.g. 2005, 2006), I want to suggest that a '*corporate* food regime' is a dominant feature of today's food system. The notion of a 'food regime' was originally intended to capture how the global food system up to 1945 was largely arranged and dominated by Britain, but then by the US after the Second World War (e.g. see Friedmann and McMichael, 1987; Friedmann, 1993). Now, however, we see a regime dominated by transnational corporations, including the suppliers of agricultural chemicals such as BASF AG from Germany and Du Pont from the US; farm machinery manufacturers such as John Deere from the US and Kubota Corporation from Japan; food processors like Nestlé and Unilever; and retailers such as Walmart and Tesco. These corporations – capitalist firms, albeit exceptionally large, and *trans*national in the sense that they have operations throughout the world – seek out greater market share, higher sales, returns on investment, and the accumulation of profits, year-on-year. They are powerful and are intent on creating a food system that works for them. And so long as they go some way toward answering the proletarian food question, it seems as if few governments around the world are willing – or capable – to challenge their power. More accurately, in fact, many governments around the world are keen to work *with* these corporations, via a tacit agreement to cede power and influence to the corporate sector and generate as much stability as possible, thereby ensuring conditions that allow investments to pay off, profits to be made, and possible challenges to systemic stability to be minimized. The outcome is a global foodscape moulded by and for the corporate sector. Eating in the contemporary period is intimately wrapped up with this corporate food regime, as I try to demonstrate in the following chapters.

Thus, in trying to understand how capitalism meets up with and shapes the global foodscapes I want to chart in this book, I think we need to keep in mind these dynamic interactions between governments, capitalist firms, and agricultural sectors. The answer to the proletarian food question has led to the emergence and consolidation of a corporate food regime. The foodscapes we see around us – the industrial-style, mechanized farm, as well as the supermarket or the chain restaurant – pivot on the actions of capitalist firms to such an extent that we might even struggle to imagine a different arrangement, never mind create one. The result is a global foodscape constituted and reproduced by all of the more local action that occurs on farms, in the factories processing food, or in supermarkets and restaurants. We are involved in all this, even if our individual actions make tiny dents in the scheme of things.

Contested global foodscapes

If the first contextual remark considered capitalism and then the emergent corporate food regime, my second is about how the global foodscapes we see around us are marked by some problematic and promising developments. In part, this is concerned with how diverse forms of *oppression* emerge from food production, distribution, and consumption. But it is also about the way numerous people around the world are *resisting* and presenting alternatives to dominant ideas and practices. Oppression and resistance: I think we do well to focus on the interplay between them. My point is a general one. The making of spaces and places is always a messy affair. We might like imagining that the world is produced via relatively smooth processes – firms getting what they want, or citizens using democratic processes to yield policy change – but these are the exception rather than the rule. More often, lots of different people chime in with their opinions, obstruct a political process, or challenge the creation of new entities, such as a new commercial zone in a city or a law designed to support or harm a specific population. At work in all of this messiness are relations of power, with some actors capable of getting what they want and others likely to lose out. Clearly, in so far as they have a history of appealing to governments to create the sort of infrastructure they need to produce and make profits, capitalist firms constitute one key set of actors here. But consumers, workers, women, or migrants also take part, offer their opinion, and shape the nature of public debate. It is little surprise, therefore, that the making of spaces and places – the making of geographies – is often heavily contested. In thinking about the making of our contemporary food system, then, it seems to me to be important that we grapple with how global foodscapes produce serious problems, even injustices, as well as calls for change. At issue are a range of processes and practices, as well as people and their diverse projects that inevitably have knock-on effects for others and therefore lead to questions, debate, argument, and struggles. It is this interaction we need to consider; this messy way in which intent and power come together in global foodscapes.

Oppression matters

In one sense, it might seem like an odd idea to suggest that the food system generates oppression. After all, food is something that gives us energy. It *empowers* us to grow and live. Moreover, if you have money today, you probably have access to foodstuffs, restaurants, recipes, and ideas and freedoms that might have seemed unimaginable until recently. In many ways, no doubt, the global food economy today is a glorious thing: food hygiene standards to give even the most anxious people confidence to eat out; TV chefs delivering wonderful new recipes; internet sites and smartphone apps helping us to work out what to cook; and commodities such as quinoa from Peru or year-round avocados for those living well beyond the tropics. Yet, without question, whenever we look at how food is produced, distributed, and even consumed we can find a range of practices and outcomes that might make most of us choke on the food we've just enjoyed eating. The food system is simultaneously a great example of human achievement and a testimony to just how awful life can be on this planet.

To grasp the problematic side of the food system, I draw upon the work of Iris Young (1990), the late political philosopher (see also Guthman, 2014). Young's interest was in understanding how social groups – such as women or blacks – suffer oppression via systemic processes that grant privileges to some and not to others. Her point was to highlight the wrongs of contemporary society and then to imagine how democratic processes might find ways to minimize or overcome those wrongs (see also Young, 2000). Part of Young's framework regarding oppression was to suggest that it had 'five faces': powerlessness, violence, marginalization, exploitation, and cultural imperialism. The general point is that any individual or social group can be subject to one or many forms of oppression, although perhaps at any one moment a particular form will matter most. Young's conceptualization of these 'five faces' asks us to be sensitive to a diverse range of wrongs: not just injustices that occur in the workplace or at home, but also in the public sphere or in more subtle, sometimes unspoken ways. I find this to be an extremely useful way of thinking about the world today and one that is highly applicable when we consider global foodscapes, as I now explain.

To begin, Young's concept of *powerlessness* is about the experience of not being able to influence key decisions that shape your life. This is not to suggest that an individual or a member of a social group has no capacity to shape anything. Nor does it mean that they are powerless regarding all spheres of their life. In short, someone can suffer powerlessness, but this cannot be taken to mean they are *always* powerless. Rather, the point is that many people will experience a feeling of powerlessness when decisions are made that they cannot influence but which have a major impact on their life. Consider a topical issue as I write this in early 2016: the development of the Trans-Pacific Partnership (TPP), a trade deal negotiated by trade officials, bureaucrats, and lawyers – with input from various industries – from 12 countries in the Pacific region of the world (e.g. see George, 2015; *New York Times*, 2015). Although it will have far-reaching consequences for everyone who lives in those 12 countries – not least food producers and consumers – the negotiations have been held in secret, notwithstanding the occasional leak to the media. This secrecy, this exclusion, has upset many people, especially members of civil society groups who have wanted to influence the process but have been kept at arm's length. In part, what makes this experience of powerlessness all the more hurtful is that it reproduces what has happened in the past: as I will discuss in Chapter 1, the World Trade Organization is based on a critically important trade agreement that fundamentally shapes foodscapes, but its creation and subsequent re-negotiation has consistently blocked efforts by civil society groups to get involved. This sort of powerlessness matters; we do well to give it our attention.

A second face of oppression is *violence*, which Young argued was a form of oppression, even if it was only threatened and not actually enacted. Violence is something many people experience on a daily basis, whether in the patriarchal household in which a father strikes his children or wife, or in regions where a US military drone patrols the skies and fires missiles at suspected 'enemy combatants'. Clearly, vast numbers of people across the world suffer this form of oppression. It is also a key feature of the contemporary food system. There are the cases of peasants lacking sufficient land who might seek to take over under-used land owned by absent landowners but receive only threats, beatings, and killings. Then there is the violence enacted against the billions of animals whose lives are ended so we can eat meat. Violence cannot be ignored when we explore foodscapes.

A third form of oppression is *marginalization*. For Young, marginalization is about the expulsion of people from useful participation in social life. It is a pervasive but under-recognized form of oppression, often because those who suffer from it do not have sufficient voice, or a venue in which to speak about it. Anyone excluded from useful participation is more likely to suffer deprivation or other forms of oppression such as violence. I think we can come across marginalization throughout the global foodscape, for example when we recognize that expanding the frontiers of capitalist 'northern' agriculture requires removing peasants from the land, leaving them little option but to move to nearby urban areas and join the ranks of those who need but cannot find stable work (see Denning, 2010).

Exploitation is the fourth form of oppression. In Young's view, exploitation is a process that transfers the results of labour of one person or group to another. In terms of the food economy – and as I noted earlier when I discussed the importance of capitalism – exploitation is fundamentally important for our purposes. We might dwell upon what happens in research laboratories and field sites; on farms; in food processing factories; or in supermarkets and restaurants. At issue here is the drive to accumulate profits, which exerts price pressure on whoever supplies capitalist firms, including workers who supply the necessary labour. Clearly, given the centrality of capitalist firms to the making of the global foodscape, exploitation is a key form of oppression for us to consider.

Finally, there is Young's fifth form of oppression, *cultural imperialism*, which is about the ways in which the makers of dominant meanings and views render those holding other views as invisible, lacking, and ultimately less important. We might consider here how the makers of 'miracle' seeds talk down and devalue traditional practices. Or how food processors talk up the importance of their presence in the food economy by emphasizing their role in answering the proletarian food question. Similarly, supermarkets promote a culture of over-consumption that produces vast piles of unused food discarded in dumpsters rather than given away to those who desperately need it. Young's concept of cultural imperialism helps us grasp these dynamics.

Five faces of oppression, then. But note that, while I will discuss what I feel are suitable cases to consider as we move through the life of food, these are not the only experiences of oppression that matter. There are many more examples out there that do not appear in this book. Consider how migrant Central American workers – many of whom take a dangerous route to cross Mexico and move into the US – experience the food system as oppressive as they pick tomatoes or oranges on US farms in Texas, Florida, or California, or work long hours for irregular and low pay in slaughterhouses (e.g. see

Mitchell, 2013). Or, imagine all of the Filipino or Indian seafarers who spend months and even years at a time away from their families to work on the refrigerated ships that move food products across the world to satisfy a rich-world year-round demand for perfectly ripe tropical fruit. And then there are peasant producers supplying coffee to global brands such as Starbucks. They might feel powerless, exploited, and marginalized all at the same time. Oppression in the life of food is about much more than what I have chosen to cover here.

From oppression to resistance

The global foodscapes at issue in this book are heavily contested. As such, it goes without saying that diverse forms of resistance oppose and challenge oppressive practices. I think this is another key aspect for us to grasp. The making of society – of the world around us – reflects diverse processes, including ones that stem from the urge to dominate, as well as the desire and compulsion to resist and create something better. There *are* actors pushing to configure the world in ways that just suit *them*, for example by making a certain type of food economy, a certain type of world in which accumulation is reproduced, or one in which domination is reproduced. But then there are *other*, subaltern or alternative ideas and practices that people draw upon and enact to create worlds that are less oppressive, more just, or more democratic. In my view, it is helpful to imagine these actions – the development of ideas and then the practices that emerge from them – as part of what we might refer to as 'repertoires of resistance' that social actors can use to oppose and work against oppression. Given that Young highlighted five faces of oppression, I want to emphasize five different forms of resistance that challenge oppression.[1] The five forms of resistance I focus on are what I refer to as imagining alternatives, occupy, going against normalized practices, striking, and boycotting.

I argue the act of *imagining alternatives* is a form of resistance in so far as it entails thinking creatively about how oppression might be undone. Rather than offering criticism alone, as many opponents of the food system might do, the task is to imagine and present ways of undoing wrongs and introducing more just alternatives. Simply imagining alternatives – even though it is rarely simple – is a significant part of trying to generate something better. It is an important element of working out what the global food economy (or any other structure) might be like were its parameters, properties, and practices made differently.

The second form of resistance is about what I refer to as *occupy*, that is, to go into the street, take over a public space, or transgress norms about who belongs in what kinds of space. To occupy is to challenge. Occupy has been discussed quite a lot lately in the context of what occurred on Tahrir Square in Egypt in 2011, where hundreds of thousands of protesters took over the square as a way to challenge the dictatorship (Bayat, 2013). And then there was the Occupy Wall Street movement (Harvey, 2012), which overlaps with other campaigns around the world such as in Taksim Gezi Park in Istanbul, Turkey (e.g. see Batuman, 2015). With respect to foodscapes, though, occupy has a longer lineage in so far as it has been used to call attention to hunger, landlessness, and inequality. Indeed, in diverse contexts, but most notably in Brazil in the last decade or so, land occupations have been used to make political points about the need to change how land is distributed as well as to secure a space in which to grow food, build a house, and live (e.g. see Wolford, 2004). In turn, occupying opens up opportunities to imagine alternatives, develop new ideas and practices, and create better social realities.

The third type of resistance is about *going against normalized practices*. When a viewpoint or practice becomes dominant, it can become normalized; that is, it can become viewed and accepted by many people as the way things should be done. Seeking to do something different therefore becomes an act of resistance against a dominant tendency. Farming the land in an organic way, without pesticides or synthetic fertilizer, is a good example to consider here (Guthman, 2004); likewise vegetarianism or veganism (e.g. see Weis, 2013). Also important, though, are the actions of some food firms. This might strike you as a bit odd; but in so far as there are firms creating products that contain or lack certain ingredients, or were made using particular processes that might have been replaced by 'more efficient' procedures, we need to recognize that they might be going against normalized practices – and as such, that their actions are forms of resistance to something their owners or managers perceive to be problematic. In response to growing awareness of how food is produced, for example, more and more people are demanding healthier or more ethical products, and some firms – big and small – are responding. Clearly, these firms are reproducing the capitalist system, at the same time as they challenge ingrained practices within their specific sectors. But in paying attention to them, what I think we grasp is that 'resistance' is not only about challenging capitalism. We have to see people or firms who go against the grain, who seek to do something different, as engaged in resistance in so far as they refuse to conform to normalized practices.

More specifically about resistance against the oppressive side of the capitalist economy is the fourth form: *the strike*. Strikes can be extremely powerful forms of resistance, especially against exploitation, but not only this form of oppression (for example, workers might strike to oppose discrimination and marginalization in the workplace). Going on strike – refusing to work for a period of time – is a brave move. It means doing without pay for the duration of the strike, and striking workers run the risk of retaliation from their employers. From the perspective of employers, of course, a day without workers is a day without production, hence a day without profits. Strikes – or just the threat of one – are therefore remarkably effective ways for workers to have their complaints heard and perhaps win improved pay or conditions. With regard to foodscapes today, I cannot think of a better moment to consider it than when we think about the sale of food. What really stands out is the case of fast-food workers in the US who have launched a campaign of strike action to protest against low pay. This movement shines a light on the role of exploitation in answering the proletarian food question; and has helped spark a broader debate about some of the ways structural inequality in the contemporary economy overlaps with race, patriarchy, and class. For our purposes, then, I consider strikes against low wages in fast-food restaurants to emphasize how key patterns and arrangements underpinning the contemporary global foodscape are contested in critical ways.

The fifth and final example of resistance I discuss is the act of *boycotting*. To boycott is to take a stance on an issue by deciding not to associate with, go near, or purchase from a particular entity. Consumer boycotts are quite common today. They might target particular products, say from a firm producing clothing in especially unsafe factories, or from firms producing in a certain country, such as Israel. I think boycotts deserve attention because they can be extremely effective ways to shape how the global foodscape emerges. They also help call attention to some of the ways food gets to our kitchens, onto our tables, and into our stomachs. Boycotts stem from new mentalities and sensibilities about the food system. As such, they offer us important lessons about how dominant ideas about food are subverted and opposed by diverse people around the world.

Of course, and as you might have concluded when you read about these five forms of resistance, they are not the only five we can imagine. There is, for example, the case of migrant farm workers striking for better working conditions; the efforts of school officials going against normalized practices to change what their students eat at lunch; or the move to tax sweetened drinks and junk food in places such as Mexico. I have chosen to focus on just five examples of resistance: your own research might reveal details of others. Further, none of the five forms I introduced above should be seen as existing in isolation from other ways of resisting. A consumer boycott might aim to support some striking workers; imagining alternatives can draw on the experience of going against normalized practices; and so on. We will do well to bear in mind the interconnectedness and the overlaps that exist between these five types of resistance. And it is important to note that practices of resistance are rarely straightforward actions from which we can easily draw conclusions about the more general interests or objectives of those involved. An individual might go on strike against an employer but engage in oppressive actions in other spheres of life; one might imagine alternatives or go against normalized practices but reproduce oppressive outcomes in other ways. In other words, then, while practices of resistance do go a long way towards making the global foodscape, we should carefully develop our consideration of them. My point is that resistance shapes what we see around us; but whether any particular example is ever going to have fundamental influence over the whole food system, or whether it contradicts other practices of resistance, is always an open question. In this sense, we must also recognize that, if resistance is one thing, actually seeing change occur is quite another. The intent and the *effectiveness* of any particular form of resistance is always up for question, not least because seeing real change in the nature of structures that have deep social roots entails long, drawn-out processes involving new laws, regulations, and numerous new practices taking shape. Nevertheless, problematic features of our world are unlikely to change on their own: someone needs to push. There is value in noting where resistance occurs, what form it takes, and how it unfolds. In arriving at an understanding of our contested global foodscape, we do well to examine practices of resistance.

On the 'life of food'

So far in this Introduction, I have remarked on the importance of capitalism and then on the need to dwell upon the interplay between five forms of oppression and resistance. I think it is useful, and in an ethical sense important, to note how our food system is contested. The corporate food regime is maintained institutionally by government policies and trade rules that boost the positions of agricultural chemical firms, food processors, and food retailers; but all along the way there are alternative voices, calls for different practices, and moves to challenge corporate control over the food system. So this brings me to my final set of introductory comments. At issue is how I organize the rest of the text.

Shaping my thinking in putting this book together is something Michael Carolan (2012) has stated in setting the scene for his book *Sociology of Food and Agriculture*. Carolan called attention to the rather blasé approach many people take towards food. For example, we know how supermarkets work and we might know how to navigate through a drive-thru restaurant, but lots of us have little clue about where the food products we eat come from; how they are produced or brought to the supermarket. Thus, he says, 'we don't think, *really* think, about the life of food prior to its arrival at our nearest supermarket' (Carolan, 2012: 2). There is a need for some caution here, because what you or I might judge to be 'really thinking' about something is always up for critique. Even so, I like Carolan's idea of the 'life of food' because it counters the temptation to rely on the related conceptualization that food has economic value and moves along a 'commodity chain' (e.g. see Pritchard, 2013). Of course, food *is* wrapped up in notions of economic value: it is sold by a peasant or a farmer for one price and bought by consumers for another; money changes hands. Further, if we think about how food gets processed, or about how value is added to it, we have to recognize the centrality of this economic dimension. Food *is* a commodity today, no doubt: crops are bought and sold, exchanged on international markets, moved across oceans and seas, and treated like most other commodities when they are milled or baked. The factory making pizzas to be sold frozen, or the restaurant selling pizzas fresh, must pay for labour, electricity, flour, oil, cheese, and other ingredients.

But while there can be no doubt that food has economic value in a commodity chain, food is unquestionably much more than all this. Just as Carolan's statement indicates, I think we can say that food has a life: there is gestation, growth, maturity, and an end, regardless of the social circumstances within which it is produced, distributed, and consumed. Moreover, food is what gives *us* life: it nourishes us, gives us strength. And we humans often take immense pleasure in cooking, eating, and sharing food. We learn from and teach using food. It is part of family life; part of our social lives. Given all of this, I find it hard to see food as just a commodity, or as simply an issue of value. In this book, therefore, I want to explore, consider, and discuss what happens to food as its life unfolds. And to do this, I divide the life of food into five stages, an arrangement which fits with the schema I introduced earlier regarding oppression and resistance.

As I discuss in more detail in Chapter 1, I suggest that the life of food begins upstream of the farm in research laboratories, corporate headquarters, legislatures, and board rooms the world over. A lot of action takes place in this upstream world which shapes what happens later on, further downstream as it were. In part, then, we need to develop a sense for the way capitalist firms look to the farm, anticipating what will be needed, and developing a capacity to supply and sell goods and services to farmers. The key point is that a vast economy has emerged to feed into and make money from the necessity of food production. The proletarian food question demands an answer and the 'solution' is to a large extent about mobilizing suppliers and other brokers – banks, lenders, and traders – to play their part. However, the world upstream of the farm is not only about firms. In numerous critical ways, the state – the agencies and institutions run by governments – intervenes to shape what happens on the land, for example by offering farmers financial support or by developing trade rules with other states that affect what farmers produce. Agriculture is governed by the national state as well as multilateral institutions such as the World Trade Organization, which shapes agricultural trade and the market for food in far-reaching ways.

In Chapter 2, I consider agricultural production in the heartlands of 'northern agriculture' and beyond. I discuss northern agriculture's input-dependence, some of the environmental questions it generates, and then how its geographical frontiers are expanding into areas where peasant-based, small-scale subsistence agriculture persists, albeit often under increasingly difficult conditions. I review some prominent developments affecting these latter food producers, consider the so-called 'yield gap' between northern and peasant agriculture which is shaping political interventions in this second stage in the life of food, and discuss how 'land grabs' are shaping processes that might have enormous potential implications for billions of people living in rural areas the world over.

Chapter 3 turns to what happens in the third stage in the life of food: when it is processed and transformed, especially by capitalist firms such as General Mills, Nestlé, and Kraft. These giants, as well as thousands of smaller companies, are all heavily invested in using diverse skills and numerous innovations to transform agricultural produce, add value, and pursue sales, market share, and profit. Like the producers of seeds or chemicals upstream of the farm, many of these firms are globalized, operating bakeries and factories all over the world. They are tied into transnational networks that bring food to the other side of the world, shaping diets in distant places, and moulding regulatory systems that have been developed by governments to oversee the food we eat. I explore some noteworthy aspects of all this action. I discuss food chemistry innovation, the development of so-called 'ultra-processed' food, and examine some of the logics driving the food giants to create markets for what are undoubtedly inventive, but yet in many instances extremely unhealthy, products. In short, I focus on the rather nasty side of food processing: on how its role in answering the proletarian food question has created problems as well as solutions. But there is another side of the story to note here. Especially in recent years, some food processors have filled the emerging niche for more nutritious, more ethical products and this, in turn, means we need to look beyond what food processors *tend* to do and ask what they *might be able* to do, were they to re-focus their energies.

Chapter 4 focuses on the fourth stage in the life of food: its sale. Competition for sales can be quite intense: fast-food restaurants hustling for our custom versus the supermarket; the local coffee shop wanting us to eat there rather than the gourmet restaurant next door; the convenience store versus its competitor along the road; and so on. Sales are far from guaranteed. There are tensions. Today, for example, supermarkets all over the world are becoming the dominant venue for food sales, out-competing older street markets, smaller shops, and even drawing in customers to their coffee shops and restaurants. They use a wide range of tactics, including taking a loss on the sale of some products in an effort to lure in customers. Some make enormous profits, but even these firms accept small profit margins. They can be extremely sensitive to costs; hence supermarkets exert significant pressure on their suppliers, including the food-processing giants and farmers, near and far. Supermarkets are powerful actors shaping what happens in the life of food, although even these firms cannot feel too secure, given that customers in some areas can now order groceries on the internet via firms such as Google and Amazon (*Reuters*, 2015). Beyond supermarkets, another crucial site in this fourth stage in the life of food is the fast-food restaurant, which deserves attention because of its ubiquity and popularity, which in turn means it shapes the way we experience, understand, and imagine food today.

Finally, Chapter 5 considers the fifth and ultimate stage in the life of food: when it is eaten and enters our bodies. I discuss what I refer to as the 'open metabolic encounter' which is about the diverse, emerging, and as-yet-unknown way we relate to food. In part, this is about issues of nutrition, specifically the public health debate about obesity. More of us are overweight or obese than ever before. This is not a big deal for everyone. For example, although they stand a higher chance of experiencing weight-based discrimination, some people who are overweight don't have a problem with it, and there are now groups such as the National Association to Advance Fat Acceptance that seek to normalize a large body shape (Paarlberg, 2010: 91). In contrast, the overwhelming medical advice today is quite clear about weight: what is best for the human body is to retain a reasonable weight and size and certainly not to become obese, especially given that obesity brings with it a higher risk of the body developing type 2 diabetes. I use Chapter 5 to examine these sorts of issues, noting debates about what leads to obesity, including questions about whether the spatial organization of our societies plays a role in generating obesity. At issue, too, will be recent scientific developments that query the assumed causal chain between eating and obesity and which also call into question the possibility that obesity might be linked to environmental toxins that change how the endocrine system functions. I also consider questions about how the open metabolic encounter connects with diverse lives, which emerge when we think about the meaning of food when people prepare it and cook.

Summary

I view these five stages in the life of food as constituted by diverse foodscapes. Such foodscapes are all around us: at roadside stalls, staff canteens, vending machines in schools, and of course on farms or in factories producing ready-to-eat meals. They include the spaces and places called to mind in the preceding discussion: the sites, the people, and scenes, and the range of relations and connections that make those places and spaces, such as trade agreements, investment flows, debates about the properties of food, and emerging ideas about how to sell it. Food is present in all of these places, spaces, and relations. The making, moving, selling, and eating of food links places and spaces with the life of food, from the production of seeds and agri-chemicals all the way along to the micro-scale action inside the human body. As we travel through these five stages in the life of food – moving from seeds, chemicals, and gestation, to processing, retail, digestion, and beyond – we need to grasp that diverse, sometimes overlapping, potentially connected places, spaces, and relations are consistently generated, reproduced, as well as altered and re-made anew. In other words, around us (and made by us) is a *highly inconstant* foodscape. Firms experiment with new recipes, entrepreneurs try out their restaurant ideas, food fashions change, and farmers change how they produce food on the land. In addition, there is a constant 're-templating' (Le Heron, 2013: 54) of where food is consumed, as populations move from countryside to city or from inner cities to suburbs and beyond. These sorts of inconstancies are features of capitalism more generally (see Cox 2002: 257–258) and they create all sorts of tensions that inevitably involve political debate and intervention. The study of foodscapes is rarely dull.

Making all this inconstancy possible, meanwhile, is also a degree of background stability – a degree of fixity – that reflects the way some capitalist firms have invested so extensively in becoming established players in the production of food and the global food economy. BASF AG, John Deere, Monsanto, Nestlé, Unilever, McDonald's, and Walmart: these and other firms like them own a vast range of assets distributed across numerous places. They are therefore determined to see those investments pay off over the short- and long-term; hence these sorts of firms have worked to create structures and mechanisms that provide them with a degree of security and protection. As I have discussed earlier on in this chapter, the resulting corporate food regime gives today's global food economy a peculiar and in many senses an oppressive form.

Finally, let me note here that my aim is to take a broad view of food and eating; one that can spend time thinking about some prominent processes and patterns shaping food production and consumption. As such, this book presents a selective rather than a comprehensive examination of prominent and emerging issues. My hope is that this book can go some way toward illuminating how our world is emerging. Food matters in all of this – and enormously so. What is taking shape around us deserves our attention.

Note

1 Perhaps I should note something here: I am not suggesting the world is arranged in some sort of straightforward five-dimensional way. Rather, my aim is merely an organizational one that brings a degree of symmetry to the following text. Even so, I think the five forms of resistance I consider certainly stand out as particularly important within a broader set of resistance practices.

References

Batuman, B. (2015) 'Everywhere is Taksim': The Politics of Public Space from Nation-Building to Neoliberal Islamism and Beyond. *Journal of Urban History*, 41(5), 881–907.

Bayat, A. (2013) Revolution in Bad Times. *New Left Review*, 80 (March–April), 47–60.

Brembeck, H., Johansson, B., Bergstrom, K., *et al.* (2013) Exploring children's foodscapes. *Children's Geographies*, 11(1), 74–88.

Carolan, M. (2012) *The Sociology of Food and Agriculture*. Abingdon: Earthscan.

Cox, K.R. (2002) *Political Geography: Territory, State, and Society*. Malden, MA: Blackwell.

Denning, M. (2010) Wageless Life. *New Left Review*, 66 (November–December), 79–97.

Food and Agriculture Organization. (2015) *The State of Food Insecurity in the World 2015: International Hunger Targets: Taking Stock of Uneven Progress* [online]. Rome: Food and Agriculture Organization of the United Nations. Available at: www.fao.org/publications/card/en/c/c2cda20d-ebeb-4467-8a94-038087fe0f6e/ (accessed 22 January 2016).

Friedmann, H. (1993) The Political Economy of Food: A Global Crisis. *New Left Review*, 197 (January–February), 29–57.

Friedmann, H. and McMichael, P. (1987) Agriculture and the State System: The Rise and Fall of National Agricultures, 1870 to the Present. *Sociologia Ruralis*, 29(2), 93–117.

George, S. (2015) *Shadow Sovereigns: How Global Corporations are Seizing Power*. Cambridge: Polity.

Goodman, M.K., Maye, D., and Holloway, L. (2010) Ethical foodscapes? Premises, Promises, and Possibilities. *Environment and Planning A*, 42(8), 1782–1796.

Guthman, J. (2004) *Agrarian Dreams: The Paradox of Organic Farming in California*. Berkeley, CA: University of California Press.

Guthman, J. (2014) Doing Justice to Bodies? Reflections on Food Justice, Race, and Biology. *Antipode*, 46(5), 1153–1171.

Harvey, D. (2012) *Rebel Cities: From the Right to the City to the Urban Revolution*. New York, NY: Verso.

Le Heron, R. (2013) Rethinking the Economic and Social History of Agriculture and Food through the Lens of Food Choice. In: Murcott, A., Belasco, W., and Jackson, P. (eds), *The Handbook of Food Research*. London: Bloomsbury, pp. 50–68.

McMichael, P. (2005) Global Development and the Corporate Food Regime. *New Directions in the Sociology of Global Development: Research in Rural Sociology and Development*, 11: 269–303.

McMichael, P. (2006) Peasant Prospects in the Neoliberal Age. *New Political Economy*, 11(3), 407–418.

Mitchell, D. (2013) 'The Issue is Basically One of Race': Braceros, the Labor Process, and the Making of the Agro-Industrial Landscape of Mid-Twentieth-Century California. In: Slocum, R. and Saldanha, A. (eds), *Geographies of Race and Food: Fields, Bodies, Markets*. Farnham: Ashgate, pp. 79–96.

New York Times. (2015) Room for Debate: The Future of Trans-Pacific Trade. 6 October 2015. Available at: www.nytimes.com/roomfordebate/2015/10/06/the-future-of-trans-pacific-trade (accessed 22 January 2016).

Paarlberg, R. (2010) *Food Politics: What Everyone Needs to Know*. Oxford: Oxford University Press.

Pritchard, B. (2013) Food Chains. In: Murcott, A., Belasco, W., and Jackson, P. (eds), *The Handbook of Food Research*. London: Bloomsbury, pp. 167–176.

Reuters. (2015) Wal-Mart Expands Grocery Pickup in Battle with Amazon [online]. *Reuters*. 29 September 2015. Available at: www.reuters.com/article/us-wal-mart-stores-delivery-onlineshoppi-idUSKCN0RT19X20150929 (accessed 22 January 2016).

Weis, T. (2013) *The Ecological Hoofprint: The Global Burden of Industrial Livestock*. London: Zed Books.

Wolford, W. (2004) This Land is Ours Now: Spatial Imaginaries and the Struggle for Land in Brazil. *Annals of the Association of American Geographers*, 94(2), 409–424.

Young, I.M. (1990) *Justice and the Politics of Difference*. Princeton, NJ: Princeton University Press.

Young, I.M. (2000) *Inclusion and Democracy*. Oxford: Oxford University Press.

1 The world upstream of the farm

1.1 Introduction

Food production is certainly about what happens on the land. Yet, long before any farmland is prepared for seeding and before any buildings are erected in which chickens or pigs will be reared, an enormous range of action occurs. For example, governments and their agencies and institutions create rules, standards, and policies that shape what food producers can do. In offices, meeting rooms, legislatures, and various other spaces far from the farm, numerous actors make calculations that alter how food is produced, traded, and sold. The nature of these deliberations matters enormously for the sort of world we are living in.

In addition, upstream of the farm – often miles away from it, in places and spaces that are entirely disconnected from food production – there is a geographically extensive, powerful and global complex of enterprises churning out the seeds, feed, advice, as well as all of the various agricultural chemicals used by farmers big and small to produce the world's food. Thus, in offices, research labs, and field sites across the world, an almost unimaginable array of experiments and tests and innovations is pursued to boost farm productivity, shape the micro-scale structure of chemicals involved in the making of food, or bring existing foodstuffs into novel relations with other foodstuffs to make entirely new food products. New inventions are made, new product innovations are developed, new ways of making farm machinery imagined, and even entirely new foodstuffs conjured up. This upstream world matters.

My aim in this chapter is to explore the terrain and consider the character of this world upstream of the farm. In part, the purpose is to describe the lie of the land: who matters, what they do, and how they shape what happens later on in the life of food. And in doing so, I think an overriding narrative emerges. Specifically, I suggest that the interests and actions of governments combine with capitalist firms invested in the food economy to pursue an answer to the proletarian food question that prioritizes input-dependent, quasi-industrial 'northern' agriculture. Action here reaches all the way through the life of food, creating structures that set parameters for what can take place later on. Here – upstream of the farm – emerges a dominant viewpoint of how food should be produced: a set of ideas and practices that gives the life of food and the associated foodscapes their peculiar form in the contemporary period.

The rest of the chapter is organized as follows. I begin in Section 1.2 with a discussion of some of the ways that agriculture interacts with what I refer to as 'governance', which is a term we can use to capture the idea that governments and other institutions shape and govern social action. I discuss agricultural supports, the state's role in shaping agricultural trade, and regulatory interventions. In Section 1.3, I turn to consider the role played by capitalist firms upstream of the farm. As I will highlight, there exists a wide range of activities and a vast complex of firms upstream of the farm, producing and supplying seeds, chemicals, machinery, and advice. Among the many enterprises in this complex, a striking feature is the prominence of a relatively small number of large, powerful transnational corporations. The discussion explores how these firms aim to transform the world in their quest to secure future profits; and how governments around the world have sought to play along in this risky game.

In Section 1.4 I note some of the key tensions arising from this emerging situation and use a brief case study on the infamous, and in many respects emblematic, seed and chemical firm Monsanto to shed light on how these sorts of tensions arise. In using this case study, my idea is to provide some background to help grasp how this stage in the life of food is oppressive. Thus, in Section 1.5 I introduce the idea that corporate influence and control over decisions affecting the life of food entails the creation of powerlessness for critical experts and those deeply affected by the way our food system is organized. Finally, in Section 1.6 I ask about resistance to this corporate dominance. I identify some exciting new ideas emerging in response to the problematic way that processes and actors upstream of the farm have altered how we produce, find, use, and eat food today.

1.2 Upstream of the farm and the making of 'northern agriculture'

Northern agriculture – capitalist agriculture – has emerged over the last few decades in the rural heartlands of countries in the global north. This is an input-dependent, mechanized, quasi-industrial type of agriculture which produces enormous quantities of food and looms large over the whole global food system. From the perspective of governments across the global north – and increasingly in countries in the global south, too – ultimately only northern agricultural practices will be able to answer the proletarian food question. Alternative visions might exist of how agriculture should operate, how it might relate with the land and ecological systems and draw upon labour, but they are marginal to a dominant understanding of how food production should occur today.

In this part of the chapter, then, I dwell upon some of the ways that action upstream of the farm produces northern agriculture. My point is that, for northern agriculture to exist – and for it to have its extraordinary capacity to produce food – an upstream world has to be put in place: projects have to be dreamt up and rolled out, practices have to occur, and diverse and numerous people have to be put to work. As I will discuss, much of this action revolves around the firms that broker deals with or supply the agricultural sector. But another crucial set of issues involves what we might refer to as the 'governance' of agriculture, and so I begin with this issue.

1.2.1 The governance of agriculture

By using the term 'governance' here, I mean that agricultural production on the land is shaped not just by local or national governments, but also by multilateral governance institutions, not the least of which is the all-important World Trade Organization (WTO). I will come to the WTO soon, so let me turn now to consider some of the ways that governments affect what happens on the land.

Given the sorts of complexities contemporary society throws up, governments are more-or-less obliged to keep an eye on things in agriculture. That regulation and governance of agriculture exists is really no big deal. Nevertheless, even if we might tend to imagine a capitalist society as 'normal' and might assume that social relations framed by capitalism are just the way things are, it is notable how governments in the world's leading capitalist societies have used their agencies and institutions – in other words, they have mobilized 'the state' – to shape a food system amenable to capitalist firms. Upstream of the farm there is action, intent, planning, deliberation. All of this reaches downstream onto the farms where food production actually occurs. I think we need to view this action as held together by an overriding consensus about what sort of agricultural system is best placed to answer the proletarian food question. But to get to that conclusion, it is necessary to step back a little and take a broad view of how the state interacts with agriculture.

Undoubtedly central to the overall story here is that agricultural sectors in places where northern agriculture has emerged have enjoyed decades of extensive government support. Although not all of these supports are explicitly directed at agriculture – a decision to build a new road that passes through an agricultural zone might benefit farmers and agri-businesses, but when governments allocate money for such a project they usually have the general population and more specifically industry in mind – many certainly are, especially those that help cover some of the costs of producing food. In other words, and to introduce a controversial concept in debates about the global food economy, the state in countries such as the US and Japan, and then in Western Europe, has tended to offer agriculture 'subsidies'.

'Subsidies' typically refer to a form of support which the state extends to businesses. Farmers, shipbuilders, even car manufacturers might all conceivably receive subsidies from the state. From the point of view of governments, these supports might be acceptable if they help employers stay afloat and retain jobs; from the perspective of firms, subsidies might be the difference between surviving and going out of business, or at least between making profits or a loss in a specific time period. Wherever we find them, therefore, subsidies highlight and reflect the existence of complex political decisions about who deserves the state's support and why. The point is: governments cannot simply extend supports to everyone; there will be winners and losers.

For anyone unfamiliar with farming the land, the nature of these supports might not be too easy to comprehend. However, consider a more obvious way that the state can support agriculture: public support for food and agricultural research (Norton, 2004: 404–420). Such research, usually based in universities, might entail work on developing the productivity of crops, controlling pests, or how best to store or move food. Since 1960 publicly funded research on food and agriculture has grown by about 700 per cent, from around $5 billion to $35 billion per year (in constant 2005 prices) (Pardey *et al.*, 2014: 3). Research by the private sector has grown as well, and now totals around 45 per cent of all research funding. In total, around $60 billion was spent by governments and the private sector on food and agricultural research in 2009, and 78.3 per cent of this was spent by high-income countries such as the US, Japan, Germany, France, and the UK (Pardey *et al.*, 2014: 3). In short, research on food and agriculture is heavily dominated by action occurring in a small group of wealthy countries. This sort of spending reflects a determination on the part of these governments to develop knowledge and capacities to improve agricultural performance. The calculation is that skilled researchers in universities can use the state's support to find innovative solutions to the developing proletarian food question and the related challenges presented by population growth and changing dietary habits. So the state intervenes upstream of the farm: it spends money, appoints panels to decide how research money is invested, and generally supports (although its level of support might rise and fall from time to time) the development of a research infrastructure that can prop up agriculture.

In addition to research, of course, the state has at its disposal a wide variety of other interventions that it can use to help agriculture or to encourage farmers to change what they do. It can, for instance, provide direct payments to farmers to help them stay afloat; offer 'set-aside' payments to take land out of production; extend loans at low interest rates; promise guaranteed prices for crops if market prices fall below production costs; offer low-cost insurance policies; and extend subsidies to cover potential losses made by exporters (see Carolan, 2011: 194). And indeed exactly these sorts of instruments have been widely used. In the US, for example, the government created a variety of policies to support agriculture in response to an agricultural crisis in the 1930s, which saw many farm incomes collapse (Friedmann, 1993). Some of these actions were intended to maintain prices, such as by offering a minimum price to farmers or by paying them to remove their worst land from production and thereby reducing supply. Then there were subsidies in the form of credit that paid a portion of fertilizer, seed, or machinery costs. The idea was to protect the agricultural sector – to lend it support, to nurture it – with a view to developing its capacity to produce. Likewise, the agricultural sector across Europe has enjoyed extensive government support. Individual countries had their own policies to promote food security or provide extension, but in 1962 the European Economic Commission, which later became the European Union, introduced a wide-ranging package of supports under the name of the Common Agricultural Policy (CAP) (see Weis, 2007: 66–67). Some supports provided farmers with guaranteed prices, while others protected European producers from imports (Friedmann, 1993). Initially these were 'productivist' supports that aimed to boost output; later they became post-productivist, in so far as the desired outputs were less about crops and more about keeping farmers on the land, even if they were not working at their full capacity (e.g. see Marsden, 2013). Supports continue to reach agriculture. For example, in the 49 countries examined in a recent Organisation for Economic Co-operation and Development (OECD) Agricultural Policy Monitoring and Evaluation report (OECD, 2015), agricultural producers received a total annual average of $601 billion in the years 2012–2014. How this money is spent varies from one place to another: some governments focus on price supports; others on reducing the costs of inputs or credit; while others still use direct payments to help cover incomes or to encourage farmers to adopt more environmentally sensitive operations (OECD, 2015: 7). The key point, then, is that rich-world or 'global north' governments intervene and extensively support their agricultural sectors.

For many farmers operating on a relatively small scale, receiving any payments or support from the state will always be welcome, and may even be what helps keep the farm going. In this regard it is worth noting here that policies to support farmers covered 18 per cent of gross farm receipts in OECD countries in 2014 (OECD, 2015). Yet agricultural subsidies do not only support small-scale farmers. In fact, in the US and Europe it is the larger farmers as well as some agri-businesses that benefit the most from subsidies. Between 1995 and 2009, for example, the US government distributed $24.5 billion to around 1.76 million farms, but another $186.5 billion to just 460 000 of the wealthiest farmers (Carolan, 2011: 193): that is, 88 per cent of the subsidies went to just 20 per cent of farms. Likewise, European subsidies are heavily skewed towards the richer farmers, with the wealthiest 20 per cent receiving 80 per cent of all subsidies there (Paarlberg, 2010: 98). Further, in both places there are numerous agricultural firms receiving supports – not least Riceland Foods Inc. in the US, which received

$554 million between 1999 and 2009, and Campina, a Dutch dairy cooperative which the EU gave €1.6 billion between 1997 and 2009 (Carolan, 2011: 193–194).

So subsidies are important. And they are controversial. In part, the controversy might be local to the countries where these supports are extended to agriculture. There may be some farmers operating at a relatively small scale who would like to see more of the money heading their way. Then there will be claims that other sectors of society have a greater need than farmers for government support: most owners of businesses with no interest whatsoever in agriculture would be delighted to see subsidies reach them. But the controversy over subsidies is also about how they play a role in shaping the overall global food economy. And to grasp how, we need to move on and consider how the WTO has influenced the governance of agriculture.

1.2.2 Agriculture and the WTO

In thinking about the importance of what happens upstream of the farm, it is crucial to grasp that we live in a world of quite significant agricultural trade flows. Of course, we might implicitly know this, especially if we regularly consume food or drinks that were produced on the other side of the world. Many British people, for instance, consider bananas to be a mainstream part of their diet, although no bananas are grown anywhere near them. Likewise, Italians and Swedes consider themselves to have relatively refined coffee cultures, but coffee is not produced in either of these two countries. A world of trade flows, then: it might just seem like an obvious, taken-for-granted, aspect of life.

However, few of us will know that the total value of exported agricultural goods in 2013 was $1.745 trillion (World Trade Organization, 2014: 61, 66; Table 1.1). Of this amount, $963 billion (53.4 per cent) was exported

Table 1.1 Agricultural exports from world regions, 2013 ($ billions)

	Total exports	*Biggest export market (and value thereof, $ billion)*	*Percentage exported beyond home region*
Europe	705	Asia (52)	23.8
Asia	431	Europe (48)	35.0
North America	266	Asia (102)	61.6
South and Central America	216	Asia (67)	83.3
Commonwealth of Independent States	67	Europe (17)	65.7
Africa	60	Europe (20)	71.2
Middle East	30	Asia (4)	40.0

Source: adapted from World Trade Organization (2014, Table II.13).

within the same general region; that is, and for example, European producers mainly rely on other European markets. But there is significant variation in the extent to which agricultural producers depend on distant markets. South and Central America, for example, overwhelmingly relies on exports to Asia, as well as North America and Europe. Likewise, North American producers ship 61.6 per cent of their exports to distant markets, especially Asia.

All of these movements – and the planning and coordination that occurs to make them possible – are shaped and governed by a mix of national government policies, on the one hand, and then by a multilateral agreement between member countries of the WTO, on the other. In effect, there is what we might imagine as a 'governance architecture' overseeing agricultural trade. This architecture is fundamental to the way agricultural trade occurs today. And it is heavily contested. To grasp why and how, consider what agricultural trade policy could be like in the absence of any sort of multilateral agreement.

Each individual country's government *could* decide to just work out their own policies to shape agricultural trade flows.[1] A government could set quotas or limits on the amount of sugar or rice that can be imported, decide to help exporters by subsidizing the cost of shipping wheat abroad, or erect tariffs – a charge on imports to bring their price up to a certain level – to strike a balance between allowing imports while not harming local producers. Of course, given that policies such as these will have far-reaching effects on all sorts of people, not least those involved in food production, they will usually entail vigorous debates between different groups in the country: maybe with small-scale subsistence farmers calling on the government to limit imports and protect their interests, while larger-scale commercial farmers ask the government to open up new export markets. Similarly, whatever policy one country decides to enact will have effects on other countries and so negotiations and debate will take place that can lead to policy adjustments, trade-offs, maybe even bilateral deals (i.e. deals between two countries). Regardless of specific debates that might emerge from this sort of scenario, however, the point here is that the governance of agriculture need not involve any sort of multilateral element. Trade can still take place, however fraught with tensions it might be; however imperfect it might be compared to some idealized notion of what entirely free trade – i.e. trade with no government interference – might achieve.

In contrast to this sort of scenario, what is absolutely crucial to understand about the contemporary scene is that the vast majority of countries are now members of the WTO and this has serious implications for how agricultural trade is governed. The WTO is based on the idea that trade works best if there is minimal government interference. The idea is that WTO trade rules will ultimately create a 'level playing field' by requiring that member countries reduce trade-distorting protections such as quotas, subsidies, or tariffs. By joining the WTO, a member country should gain access to new markets for its exports and this, in theory, should enable businesses in each country to find a niche in the world capitalist economy. Similarly, member countries are supposed to reduce their control over imports and this means domestic economic actors will be exposed to the most competitive products using the latest technology, say, or the most productive labour. In turn, that exposure might compel them to innovate and become more productive and competitive, which in turn should boost the overall global capitalist economy and yield economic growth and improved material conditions for all. This, at least, is the general idea that makes the WTO, if not the world it envisions, a reality.

Agriculture was brought into the WTO as a result of the Agreement on Agriculture, which introduced a framework for ensuring that individual governments would abide by multilateral trade rules affecting food production (see Weis, 2007: 128–160). This agreement emerged in the context of debates about the relative power of agricultural sectors in the US and Europe. Since at least the 1950s, farmers in the US – backed up by price supports and subsidies – produced so much food that its government looked to find ways of disposing of it abroad. Its solution to the 'state of chronic surplus' (Weis, 2007: 63–70) was US Public Law 480, which was launched in 1954 by the US government to fund 'long-term loans to developing countries at below commercial rates for the purchase of U.S. grains, mainly wheat' (Clapp, 2012: 28). A similar outcome emerged in Europe. The CAP had the initial aim of simply ensuring food security, but farmers responded so well to the supports on offer that surpluses soon emerged. The infamous European 'butter mountains' and 'wine lakes' sought to capture the idea that European consumers could not eat what farmers were producing, hence huge stocks of unsold butter or wine were accumulating. Following the example of the US, Europe created a food aid programme in the 1970s that used donor countries to dispose of surpluses, especially of dairy products, grains, and meat (Weis, 2007: 66; Clapp, 2012: 29). In effect, these policies enabled the 'dumping' of subsidized food on the world market.

Support for the WTO's Agreement on Agriculture was partly about some governments wanting trade rules that would place a limit on food dumping. But it also reflected moves from some other governments – especially the Cairns Group, which included Australia, Canada, and New Zealand, along with Argentina, Brazil, Thailand, Malaysia, and 11 others – who wanted their commercial farmers to have far greater access to export markets. As is the case today, the largest markets for agricultural sales at this time were the US and Europe, but farmers outside of those two regions struggled to make sales there, in part because market access was restricted behind quotas, subsidies, and tariffs. Finally, another crucial part of the story here was that agricultural transnational corporations (TNCs) – a group of upstream actors I discuss in more detail later on in this chapter – were intent on seeing multilateral agreements take shape that would entrench their power. As Weis (2007: 132) notes, 'they wanted to maximize their flexibility selling and sourcing within and between nations by shifting the locus of sovereignty within the global food economy, moving significant elements of regulation beyond the legislative reach of governments'. Further, they wanted to see instruments established that would protect their investments in research and development, especially the all-important innovations some TNCs were pursuing with regard to seeds.

Overall, then, the WTO's Agreement on Agriculture was about working towards the creation of a liberalized trading environment in which trade-distorting barriers would be gradually removed or at least reduced. It was about producing a framework – an architecture – that would reduce government controls and locate decision-making power within agreements between member countries. Yet, in a striking and problematic twist, rather than yielding this outcome, the WTO has ended up creating a framework that actually permits all sorts of subsidies and locks in place the dominance of rich-world, northern agriculture. The key factor here is that, although the Agreement called for an overall reduction in 'trade-distorting' subsidies, its definition of this term was 'limited' (Weis, 2007: 136) and this has had far-reaching consequences.

The most obvious trade-distorting subsidies were export subsidies and payments to farmers that rose according to how much they produced, which encouraged overproduction and led to an over-supply and falling prices. The Agreement prohibited these or – if they were labelled 'amber box' subsidies ('amber' as per a traffic-light analogy, i.e. 'go slow') – they were to be gradually scaled back. In one sense, then, the Agreement was a partial victory for free trade: some trade distortions, and therefore the overall scope for government interference, were reduced. However, the Agreement permitted numerous other forms of support, which were labelled 'green box' ('green', as per a traffic light for 'go ahead') and no restrictions were placed on what governments could spend on them. Included in the green box were supports such as 'research and extension expenditures, income supports, land set-aside payments, early retirement for farmers, deficiency payments to farmers, regional assistance programs, and crop insurance' (Clapp, 2012: 70). And so the result is that the WTO has actually allowed the world's richest countries – including the US, Europe, and Japan, which account for 87.5 per cent of the total global spending on green box supports (Carolan, 2011: 19) – to spend vast amounts of money on their agricultural sectors, just so long as their supports are not classified as trade-distorting. Although the Agreement does not view these supports as trade-distorting, in reality, and as Weis (2007: 137) notes, 'the impact of subsidies is much more of a grey area'. In effect, what the green box has allowed countries to do is 'hide their trade-distorting subsidies' (Carolan, 2011: 19):

> in practice, these payments still shield producers from low prices – prices that would otherwise, in a less distorted market, send signals to farmers to produce less or something else. Green box policies therefore do

> distort markets. They make farmers deaf to market signals, allowing them to continue to (over)produce and profit even when the costs of production exceed what the market is willing to bear.
>
> (Carolan, 2011: 20)

And the upshot is that, in contrast to what the discursive agenda behind the WTO's Agreement on Agriculture claimed, food dumping from the global north into the global south has not stopped. Farmers in the US, Europe, and Japan are still subsidized, and this means they can sell food abroad at a price below its production cost. Indeed, between 1997 and 2003, 37 per cent of wheat and 19.2 per cent of corn exports from the US were sold below production cost (Carolan, 2011: 21); and still today farm incomes throughout the world's richest countries are propped up by government support (OECD, 2015). Rather than the WTO limiting food dumping from the heartlands of northern agriculture into poorer countries, it has actually made it more likely. According to the Worldwatch Institute, for example, 13 countries are now 100 per cent dependent on grain imports, while another 51 are 50 per cent dependent, which reflects a pattern of growing dependence on subsidized food (*Reuters*, 2015). Of course, for those who can afford to purchase imported wheat or maize, or processed versions thereof, such food imports might very well be welcomed (although the amount of ultra-processed, low-quality food has numerous negative health consequences, as we will see in Chapter 5). But the other side to consider here is that dependence on food imports means consumers risk becoming unable to afford them if prices rise, as they did in 2008 with dramatic consequences throughout the world (e.g. see Clapp, 2012). And because these food imports negatively affect the prospects for agricultural sectors in some of the world's poorest countries, this outcome is, to a large extent, the source of the controversy about the WTO and its role in creating a governance architecture that props up northern agriculture.

There are some other noteworthy issues to consider here. One is that, in shifting some control over trade rules to the WTO, governments also lose some scope to make democratic decisions that run counter to what previous governments have agreed to. Maybe producers in their countries do gain access to new export markets – and maybe this does benefit employers and provide employment for many workers – but the trade-off is this lost room for manoeuvre. Decision-making power has been relocated away from national legislatures. The net result is that the WTO locks countries into a structure of relations with other countries and the firms that produce, move, and trade goods and services. Economic growth might be the reward, but governments have far less scope to shape what form that growth will take.

Second, with regard to food and agriculture, imbalances regarding subsidies within the WTO mean that all governments are now under pressure to provide domestic supports – green box, of course – that reinforce the expansion of northern agricultural practices: input-dependence, consolidation, and the general drift towards greater commercialization of agriculture. Again, this might be fine if we only view agriculture as an economic practice – as something that should only be about profits and accumulation. But for many people today – not least the 2.5 billion people who still rely on eating the food they produce (World Bank, 2007: 3) – agriculture has a rather different meaning: it is about eating, surviving, even if some of what they produce will be sold on the market. Or, for many small-scale farmers in the heartlands of northern agriculture, it is about maintaining a livelihood while also protecting the land and having something to pass on to their children. Not just profit. Not just accumulation.

Thus, in enacting rules and generating a structure that values and supports commercial, capitalist northern agriculture, the WTO and the overall governance architecture of which it is a part closes down the sorts of options open to all of us as citizens and indeed voters: ultimately, should we decide to elect governments that make commitments to pursue an alternative agenda with respect to agriculture or other sectors of the economy, we will need to confront WTO trade rules and the reduced scope they allow for making decisions that run against market rule. The result is a governance architecture promoting an input-dependent agricultural system to answer the proletarian food question. Overall, then, upstream of the farm – via deliberations between officials, bureaucrats, politicians, as well as lawyers and lobbyists in offices, meeting rooms, and legislative chambers all over the world – a wide range of decisions have been made that affect what happens on the land. As for whom all this benefits: this is the moment to consider the capitalist firms operating upstream of the farm.

1.3 Capitalist firms and agriculture

There are plenty of capitalist firms taking a keen interest in agriculture. At issue are profits that might be made by selling goods or services to farmers, lending them money, or trading their produce. Upstream of the farm, therefore, there is a vast complex of capitalist firms looking to see what farmers will do. They anticipate planting, the tending of crops, and harvest time. They look to make money from selling services or goods, or trading what leaves the farm. In effect, they pursue their own interests by reaching onto the farm, as well as further downstream, and their actions alter what happens and what *can* happen to farmers, and all others besides. What I want to do here is consider who these firms are and what they do. So far as I see it, there are two main sets of firms to consider.

1.3.1 The brokers

The first set of firms consists of what I think we can refer to as the 'brokers'. Included here are the trading companies that buy and move agricultural produce around the world. Archer Daniels Midland, Bunge, and Cargill are three of the largest examples. They monitor the world's markets to identify where surpluses and deficits might emerge and move to ensure buyers can get the produce they need, and sellers can get their products off to market (Clapp, 2012: 91–93). In many senses, this is about a rather raw economic logic of buying low and selling high; and a global agricultural market such as we find today needs intermediaries to do the job Bunge or Cargill perform. However, there are concerns that they abuse their market power. The top firms control up to 90 per cent of the world grain trade and move vast quantities of grain to millers and food and animal feed processors (Clapp, 2012: 98). As such, this small number of trading companies can squeeze farmers by only offering low prices for their output. Meanwhile, trading firms make significant profits. In 2014 Cargill made a profit of $1.8 billion from worldwide revenues of $134 billion (Cargill, 2015).

Another prominent group of brokers are market speculators based in cities with large financial sectors, such as Chicago or New York. They are instrumental in shaping global food prices. They receive regular updates from providers of agricultural statistics and use these data to decide whether to buy or sell agricultural produce, as well as shares in companies producing agricultural inputs, or processing or retailing food. In addition, they buy and sell 'futures' contracts, which farmers enter into with traders in advance of harvest time in the hope of guaranteeing a good price for their crop. Much of what these speculators do can be carried out in the same way with other non-agricultural goods; in many senses it makes no difference to these actors that they are dealing with food and that their actions – especially when they lead to rising prices – can make food unaffordable for millions, if not billions, of people around the world. Clearly, though, in front of a computer in a high rise on Wall Street, or on trading floors in a place such as the Chicago Board of Trade, the actions of farmers and indeed smallholders and peasants around the world are imagined, calculated, and anticipated. This 'financialization' of food (see Martin and Clapp, 2015) makes these trading sites in New York or Chicago, as far as they are from farms and the land run by peasant households, critical to the overall character of the contemporary global foodscape.

A final group interested in brokering deals with farmers focuses on lending them money. In many cases among the heartlands of northern agriculture these lenders are simply local banks, albeit usually branches of larger commercial or corporate banks with interests beyond agriculture. Loans might be small-scale and used to pay for repairs or inputs, while larger farm operations will turn to larger banks. The point: a market exists for lending to farmers, with banks keen to find new customers and happy to see farmers use debt as a way to pay for inputs such as seeds or fertilizer (see Martin and Clapp, 2015). For many farmers, of course, access to credit will often be welcomed; but wherever debts are incurred, risks necessarily emerge, not least that small farmers might lose their land if they cannot repay loans (e.g. see Patel, 2007: 46–49). Debt is always a double-edged sword: it has to be repaid with interest and for many farmers the interest will take up anywhere from a large portion to all or even more than any surplus left over once crops have been sold and other non-debt-related costs have been paid.

Beyond places such as the US or Europe, where profit-driven commercial banks are accustomed to lending to farmers, cooperatives are critical lenders; and predatory money-lenders are rarely far from the scene (e.g. see Akram-Lodhi, 2013: 45–46). Many governments are also involved in agricultural credit. Agricultural development banks, a feature of the days when governments and state officials believed they could intervene in society by lending money and guiding how 'development' would occur, still play a major role in countries throughout the world (see Table 1.2). Yet, in some countries, as a result of reforms that aimed to give the private sector a greater role in society,[2] governments have sold agricultural banks – that is, the banks have been privatized and made into profit-making commercial banks, rather than a state-owned asset – in the hope that a commercial orientation will expand access to credit for small-scale farmers. In Guatemala, for example, the agricultural bank Bandesa was closed in 1997 and Banrural created in its place. It collects savings from 1.3 million accounts, dispenses loans to 200 000 clients, and has a governance structure which claims to balance the need for profitability, on the one hand, with rural development, on the other (World Bank, 2007: 146).

1.3.2 The input suppliers

A second set of firms consists of the suppliers of services and goods – such as seeds, fertilizer, pesticides, machinery, feed, and advice – that are intended to boost farm productivity. These goods get slotted into the farm, hence they are known as 'farm inputs'. Although supplier firms come in all shapes and sizes, from the independent agricultural adviser, to the giant producer of agri-chemicals, a key feature here is the emergence of some giant supplier firms. Table 1.3 lists the ten largest firms globally, based on data from 2007

Table 1.2 Number of borrowers from agricultural development banks in selected countries

	Number of borrowers
Thailand (BAAC)	5 million
Egypt (Principal Bank for Development and Agricultural Credit)	3.5 million
Indonesia (Bank Rakyat Indonesia)	2.5 million
Pakistan (Agricultural Development Bank)	700 000
Iran (Agricultural Bank of Iran)	600 000
Nepal (Agricultural Development Bank)	400 000

Source: Food and Agriculture Organization (1999: 39).

Table 1.3 Ten largest supplier transnational corporations, ranked by total sales, 2007

		Sales	
	Home country	*Foreign*	*Total*
BASF AG	Germany	49 520	85 310
Bayer AG	Germany	24 746	47 674
Dow Chemical Company	US	35 242	53 513
EI Du Pont	US	18 101	29 378
Deere and Company	US	7894	23 999
Yara International	Norway	9939	10 430
Syngenta AG	Switzerland	9281	9794
Kubota Corporation	Japan	4146	9549
Monsanto Company	US	3718	8563
Agco Corporation	US	5654	6828

Source: adapted from UNCTAD (2009: 240).

and provided by the United Nations Conference on Trade and Development (UNCTAD) in its 2009 *World Investment Report* (UNCTAD, 2009: 240). Included in the list are suppliers of agricultural inputs such as BASF and Bayer from Germany, Dow Chemical and John Deere from the US, as well as Syngenta from Switzerland and Kubota from Japan. Figure 1.1 illustrates the sales, number of employees, and operating countries of BASF, Bayer, Dow, and Du Pont. Worth noting is that all of these firms have headquarters (HQs) in so-called 'developed world' countries: indeed, of the largest 25 firms, eight have HQs in the US, three in Germany, while Denmark, Japan, Norway, and Switzerland are each home to two of them. The world upstream of the farm – the complex of firms, their activities, and their capacity to shape the life of food – is heavily dominated by the interests, backgrounds and imagined futures of a small section of the world, mainly Europe and North America.

The combined sales of these firms, as reported in the UNCTAD report, totalled $321 billion in 2007 (UNCTAD, 2009: 255).[3] Sales outside of their home region mattered immensely for most of these firms. For those from small countries such as Norway or Switzerland, this almost goes without saying; but even among the five largest firms from the US, 57.3 per cent of all sales were foreign, whereas the average among the largest 25 firms in this sector was 61.3 per cent. In other words, these are firms that look beyond their borders for sales, for opportunities, for their futures. They have a *global* reach, backed up in some cases by alliances between firms that look for ways to secure profits by working together to produce, move, and sell their goods and services in 'input bundles'. Indeed, UNCTAD reports that 'a few very influential alliances of TNCs have emerged which span various upstream and downstream stages of respective value chains. The three most advanced alliances of this sort are alleged to be Monsanto/Cargill, ConAgra and Novartis/ADM (Archer Daniels Midland)' (UNCTAD, 2009: 152).

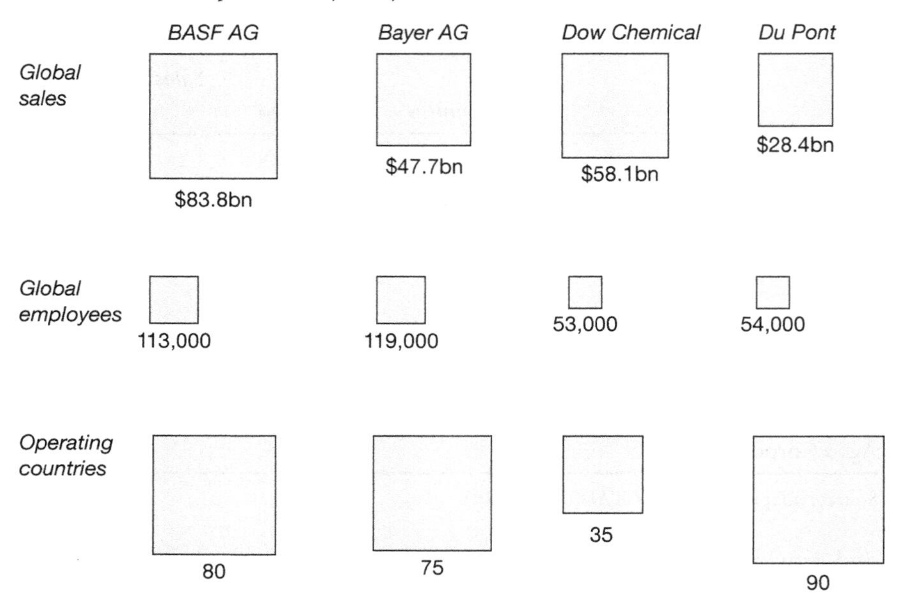

Figure 1.1 Global sales, employees, and operating countries of four agricultural chemical suppliers, 2014 (sources: BASF AG (2014), Bayer AG (2014), Dow Chemical (2014), Du Pont (2014)).

The horizons of these TNCs have expanded considerably in recent years and this has given them the urgent sense that almost the whole world can become their marketplace. In their public relations, in their websites and speeches, representatives from these firms imagine the whole globe as their marketplace, drawing in commodities and churning out products via global supply chains, and scanning the globe for new insights that might lead to further innovations. They are driven to grow, for example by purchasing competing firms in a process known as 'consolidation', i.e. firms consolidate, and become bigger, by purchasing and therefore reducing the number of competitors. These acquisition strategies, made possible by past and projected profits, as well as access to credit, have enabled the biggest supplier firms to dominate specific markets. For example,

> the global supply of proprietary seeds and agrochemicals is controlled by only a few TNCs [with] the top four seed TNCs control[ling] 53 per cent of the global proprietary seed market: the leader – Monsanto – accounts for 23 per cent of this market.
>
> (UNCTAD, 2009: 152, quoting ETC Group, 2008)

with even more striking results in particular seed markets: the top three firms account for 65 per cent of the maize seed market and half of the soybean seed market (Clapp, 2012: 105). Furthermore, in 2008 Monsanto had 9 per cent market share of the global pesticide market, with Dow Agrosciences (10 per cent), BASF (11 per cent), Bayer (19 per cent), and Syngenta (19 per cent) ahead of it. These are truly enormous and powerful firms with a global footprint, ready and able to sell their products to farmers anywhere in the world. They now occupy a central position upstream of the farm.

From the perspective of farmers, the size and reach of all these firms presents a bit of a dilemma. On the one hand, access to inputs is much needed, especially if fertilizer or pesticides can boost output and keep the farm operation going. On the other hand, though, the size, strength, and market power of these firms raises the likelihood that the price farmers will need to pay for inputs will be unnecessarily high. Concentration of market power upstream of the farm breeds dangers and risks. Regulation via oversight from the US Departments of Trade and Justice, or the European Union's Commission, might go some way to prevent abuses; but it can be extremely difficult to prove price fixing.[4]

More generally, the rise and power of supplier TNCs raises concerns that they will shape how governments and states regulate and govern agriculture. As Jean Ziegler, the first United Nations Special Rapporteur on the Right to Food, noted: 'As financially powerful lobbying groups, corporations ... exert great control over laws, policies and standards applied in their industries, which can result in looser regulation and negative impacts on health, safety, price and quality of food' (UNCTAD, 2009: 157, quoting United Nations, 2003). TNCs have sufficient resources to: send expert lawyers to trade meetings; lobby politicians contemplating new legislation; or launch media campaigns to defend their interests. For example, the large TNCs seek to shape other regulations, such as the Codex Alimentarius Commission, which is run by the FAO and the World Health Organization (WHO) to establish international food safety standards. Some civil society organizations criticize the commission for 'failing to include the participation of small producers and consumers, and being heavily influenced by the lobbying and participation of large agribusiness, food and chemical corporations' (UNCTAD, 2009: 157–158, quoting United Nations, 2003). Furthermore, some TNCs were closely involved in shaping the TRIPS Agreement (Trade Related Intellectual Property Rights Agreement), which set up a framework within the WTO for protecting intellectual property rights, not least for microorganisms and biological processes. TRIPS

> paved the way for [transnational corporations in the agricultural biotechnology industry] to globally market agricultural biotechnology products because they could be assured that their patented varieties of seeds would be legally protected from being replicated and sold without compensation to them.
>
> (Clapp, 2012: 74)

Such has been their power that the large supplier firms have been able to push for and achieve these sorts of changes in policies and practices across the world; and changes that strongly favour their interests. They lobby and argue for friendly policies and regulations via discursive and material efforts to craft a food economy that suits their interests, rather than a more careful balance of corporate, public, and environmental concerns.

Consider research again. In rich-world countries, the private sector accounts for just under half of all agricultural research and development (Pardey *et al.*, 2014: 3). Of course, profit-driven research responds to opportunities to market and sell inventions – in shorthand, it responds to wants, not needs. Some of this research activity *might* lead to new ways of addressing hunger, say; but a far more urgent concern will always be that it creates new profit streams. As such, private sector researchers focus on developing the quality and applicability of farm inputs such as agri-chemicals, drugs, and machinery that are available only to those who can afford to buy them. Corporate research and development also tends to focus on the small range of crops grown by temperate-zone commercial farmers, at the expense of other crops, such as sorghum, which is a crucial source of food in sub-Saharan Africa, where hunger rates are among the highest in the world. Further, private sector research might investigate 'moonshots' that look to the future and imagine how a rising population with a taste for meat might be satisfied, for example by addressing the so-called 'protein crunch' by converting protein from insects (see Drew and Joseph, 2012) or plankton sucked up at an enormous scale from the southern oceans into 'burgers' or similar products (Drew and Lorimer, 2011).

Connected to all this are concerns that TNCs are building on what is already known about genetically modified organisms (GMOs) by pursuing a so-called 'Gene Revolution' – as opposed to the *Green* Revolution – that aims to reach into and modify the genes of numerous species. These technologies have the potential to dramatically change the way food is produced, consumed, and understood. In recent years, for instance, a debate has raged over the development of so-called 'golden rice', which has an altered genetic make-up that makes it unusually rich in vitamin A (e.g. see Patel, 2007: 136–137). Golden rice is intended to help people who suffer from vitamin A deficiency (VAD), which affects about 140 million people each year, mainly across the tropics. The idea of growing golden rice is that it can deliver the much-needed vitamin A within a staple food. The debate is over whether golden rice is truly the best way to address VAD. For some, golden rice is exactly the sort of genetic engineering of food that science needs if the world's population is to be adequately fed in the twenty-first century (Paarlberg, 2013). But as some other commentators have noted, the real problem in a place such as Bangladesh is inequality and poverty, which limits the sort of food production and consumption that can occur, thereby leading to a restricted and vitamin-deficient diet (Patel, 2007).

More broadly, what the case of golden rice raises is the possibility that private sector agricultural research will score scientific victories that go beyond public, democratic scrutiny and control, with potentially long-lasting or even irreversible consequences for human and non-human life. It is almost impossible to control what happens once GM crops are grown, pollinated, and enter the food chain. For many scientists, precaution is the best approach; to their relief, some countries, not least Germany – a country not known to shun technological advances – have placed moratoriums on GM crops. Yet globally the area planted with GM crops has grown to reach 181 million hectares of land, or just over 10 per cent of the 1.5 billion hectares of cultivated land, with much of this in the US, where regulatory decisions in 1995 were made with 'scant public input' (Macilwain, 2015). That private, profit-driven corporations stand to gain from further growth goes without saying. From the perspective of citizens and consumers, though, while GM crops could be the cure to rising food prices on a growing planet – as corporations such as Dow Chemical and Monsanto preach – the risk is that these developments alter the genetic make-up of humans and non-humans alike, with unpredictable consequences.

Beyond research, consider that, over the last few years, agricultural insecticides, particularly those based on neonicotinoids, have come under scrutiny because they seem to kill many more pests than their manufacturers claim (Carrington, 2014). Whereas these insecticides are supposed to die away in the soil relatively soon after they have been applied, studies have shown they have a much longer half-life and this has led many specialists to raise concerns about the links between the use of insecticides and the rapid decline in the number of some insect species, especially bees. In Europe there has been a push against these insecticides, resulting in the European Commission using a European Food Safety Authority report to decide in 2013 to introduce a two-year moratorium on the clothianidin, imidacloprid, and thiamethoxam compounds used in such insecticides. This has rankled firms such as Syngenta and Bayer, who have responded by funding industry associations such as the Crop Protection Association to lobby and challenge the European Commission's decision. They argue that pesticides help maintain food production and that, without them, we would see steep declines in output and rapidly rising prices. Given the power and influence supplier TNCs can conceivably wield, there are numerous concerns that their activities or products will work primarily for their owners and shareholders, rather than for the planet as a whole.

Thus, if the above discussion demonstrates anything at all, it is that numerous profit-driven activities shape the character and form of the global food economy in striking ways. Long before anything happens on agricultural land, there are capitalist firms working to lend, speculate, and trade; supply inputs; develop new products; and lobby to shape how governments regulate and govern agriculture. Right here in the initial stage in the life of food, in the gestational phase, the global food economy is generated and shaped by private actors looking to profit – and almost at all costs. The first stage in the life of food has become a private and privatizing arena; a deeply globalized complex made up of large, powerful, profit-driven enterprises, looking to out-compete each other and secure future profits by consistently developing new product lines, all manner of new innovations, and new markets. And so this is the sort of action which supports the claim that a 'corporate food regime' (McMichael, 2005, 2006) shapes food production today. As I have discussed, governments are certainly involved in all of this. Especially in rich-world countries, they have used their powers to write policy, craft new rules, and broker trade deals to suit a type of input-dependent, quasi-industrial 'northern agriculture'. But it is also crucial to see how the corporate sector has sought to use the state to promote a regime which values wants over needs. This is almost taken for granted; assumed to be 'just the way things are', even though it has emerged because agricultural sectors throughout the rich world have received support from the state and its capacity to make critical interventions. Oddly, then, the corporate food regime is both an indicator of a world organized around and for TNCs – and so very much a 'neoliberal' world, if we take neoliberalism to refer to a contemporary emphasis on the social value of private sector involvement in society (e.g. see Harvey, 2005; Klein, 2007; Brown, 2015) – and yet a signal that the state remains a central actor in the making of that world.

Upstream of the farm, therefore, in government offices and meeting rooms – via the combination of inspections and monitoring technologies, and with the work thrown in of lawyers, trade officials, and various other brokers and agents – a whole host of work is done to create northern agriculture. Likewise, upstream of the farm there are numerous and diverse moves to roll out privatized credit; expand corporate control of agricultural trade; and create a complex of firms and public bodies producing a wide range of goods and services that make agriculture a radically different activity to what it once was, even as recently as 50 years ago. Billions of dollars of investment in private and public research laboratories occurs upstream of the farm to find new ways of treating and changing the crops, not to mention the lives of the animals, we eventually eat. There is a world of activity that reflects investment decisions, consultant and shareholder assessments of the future of food, and then the spending of vast amounts of money to make the goods and services that are sold on to farmers. And because all of these goods and services need to be paid for, and even though farm productivity might improve in the immediate or medium term, they become costs that farmers must bear, perhaps even on an increasing basis. In turn, of the money flowing into the hands of farmers globally, a growing portion must be passed back upstream to the developers and producers of all these inputs. Even if it remains incomplete, contested, and far-from-inevitably hegemonic or dominant, this corporate food regime, and the spaces and practices upstream of the farm on which it is based, is a crucial feature of the way agriculture occurs today.

Accordingly, in this sort of view, agriculture primarily should be about boosting economic growth, not about ensuring everyone can eat. Maybe this is the way things should be organized. Perhaps we should conclude that the suppliers of chemicals or machinery are a vital part of the current global foodscape: they have played a key part in our growth, the food we have eaten in our lives, and they will likely continue to play such a role in our futures. Moreover, this corporate food regime certainly does deliver food, albeit only for those who can afford to buy it. And no doubt many farmers who are plugged into the system and have the money to buy land, fertilizer, or other inputs, and can therefore make a good living, will be vocal advocates for the sort of upstream world we have been charting here. On the one hand, then, it is perhaps foolhardy to expect or wish them to disappear. In a global foodscape that takes food justice seriously, we are probably still going to need input suppliers of some sort. On the other hand, though, and as I noted earlier, many TNCs – like many of their peers elsewhere in the capitalist economy – pursue their own interests at all costs, even if this comes at the expense of people and our environments. Thus, perhaps we *will* need large enterprises to supply agricultural inputs in a global foodscape that takes food justice seriously, but at issue will be finding a democratic way of balancing their interests against the interests of humans, other species, and the environment.

1.4 Case study: the case of Monsanto

To grasp some of the complexities at work here, I now turn to consider one of the most powerful TNCs upstream of the farm: Monsanto. Established in 1901 in Missouri in the US, Monsanto is now one of the world's largest agricultural supplier firms. In 2007 it was the largest seed firm in the world, with 23 per cent market share, ahead of DuPont and Syngenta (UNCTAD, 2009). Monsanto is big. And it is extremely profitable. It made as much as $2 billion in 2008 from $11 billion revenues, or a profit rate of 18 per cent of revenues, which most firms can never dream of (Clapp, 2012: 107). It is deeply invested in maintaining the sort of corporate food regime in which it succeeds. It actively supports Crop Life and the Biotechnology Industry Organization, both of which aim to shape policy decisions regarding biotechnology across the world; and is a supporter of the International Life Science Institute, another lobbying organization with a global reach. Monsanto uses its public relations budget, its website, and other opportunities to promote the message that its products, especially its GM crops, are 'pro-poor' (Clapp, 2012: 121). And when trade deals or new arrangements are negotiated, it will frequently aim to get involved, as do many of its peers.

For example, when the terms of the WTO were worked out in the early 1990s, Monsanto was a major advocate of finding ways to ensure intellectual property over seeds and other organisms could be protected worldwide (Clapp, 2012: 106). Intellectual property rights allow innovators to create patents over their inventions, which prevent competitors from copying them without paying a hefty fee. Such patents – of which Monsanto owns 'more than 400' (Paarlberg, 2010: 130) – can last for up to 20 years, during which time the owner can effectively operate without competition. Reflecting the way the seed firms are producing for a market, much like the market for clothes or smartphones, many of these patented seeds have brand names, advertisements, and marketing campaigns.

In Monsanto's case, a key brand is its Roundup Ready seeds for crops such as alfalfa, corn, cotton, soybeans, spring canola, sugarbeets, and winter canola. These seeds have been genetically modified to allow farmers to spray the herbicide glyphosate, which Monsanto sells under the name Roundup (Paarlberg, 2010: 163), 'from emergence through flowering' (Monsanto, n.d.), thereby limiting the growth of weeds and, in turn, helping farmers to maximize yields. When farmers buy Roundup Ready seed, no matter where they are in the world, 'they get a packet and a long, densely written legal tract, in a language that would be barely comprehensible to them even if it were written in their native tongue' (Patel, 2007: 134). Farmers who use patented seeds cannot 'save and replant the seeds after harvest [which in turn means their seeds] must be bought from the company again the next season' (Paarlberg, 2010: 130). Violators will be prosecuted. Thus, the firm has used its legal authority to chase farmers – even farmers operating on a small scale and who stand little or no chance against Monsanto's massive legal war chest. In what seems to have been a landmark case, in 2013 Monsanto won a legal challenge brought to the US Supreme Court over Indiana farmer Vernon Bowman's use of Roundup Ready seeds, which he planted against Monsanto's legal terms (e.g. see Liptak, 2013). Bowman challenged those terms but lost, which now gives Monsanto the legal right to continue going after further patent infringements. Monsanto's determination to use its patent protections against small farmers rings alarm bells among critics of the food economy (e.g. on mobilizations against Monsanto, see Scoones, 2008).

It might be necessary to say here that some farmers will not complain too much about Monsanto if their seeds do indeed boost output. This is part of the dilemma we all face. The global foodscape is constituted and in many senses also directed by firms such as Monsanto. For those of us who are fortunate enough to afford food: well, much of the food we eat – and certainly its price – hinges to a large extent on investments and innovations made by these TNCs. But what the case of Monsanto highlights is the knock-on effects for agriculture, and the food system as a whole, of powerful TNCs having the scope to shape what happens upstream of the farm, as well as throughout the life of food. Monsanto has sought to protect its investments via legal means, but its size and profitability also enable it to mobilize political supporters to further bolster its position. In this regard it is vital to note that numerous former Monsanto employees and executives have found political positions in the US Food and Drug Administration and the

Environmental Protection Agency (Ferrara 1998), while some former government officials have left to take positions with Monsanto. The fear among critics about all this is that Monsanto will use these sorts of favourable ties with government agencies and departments to navigate and evade food safety regulations and environmental standards, or to encourage the government to introduce regulations that Monsanto, but not its competitors, will manage to meet, thereby stifling competition and protecting the firm's power (Weis, 2007).

Overall, then, as I think the case of Monsanto makes clear, the capacity of corporations to shape food production – a capacity made possible by the background stability TNCs look to secure via the corporate food regime – raises numerous concerns. Monsanto is in a strong position and its future profits (and all the bonuses and salaries to its top executives) depend on it managing to protect that position. Like the other big players in the world upstream of the farm, like the top ten firms noted above as well as all the other smaller firms, Monsanto needs the corporate food regime to be maintained and strengthened. And given the way political systems can be held captive by corporate donations, or by offers of favourable positions in board rooms for compliant officials, the risk is that human health or environmental consequences will come second to the all-important drive for greater market share and future profits. Monsanto's growing power reflects a broader set of developments in contemporary capitalism: as transnational corporations grow, so does their capacity to shape and even subvert the democratic process (e.g. see George, 2015). One outcome is the existence and growth of an array of corporate interests that watches, waits, anticipates, plans, and then invests in the reproduction of a food system shaped by the corporate food regime.

1.5 In what sense oppressive?

What the various supplier TNCs upstream of the farm have managed to create over the last few decades is a global food economy in which their interests are increasingly dominant, even if their interests come at the expense of human health and the environment. The undeniable general direction of change upstream of the farm has been for the giants to operate without having to pay too much attention to their critics, who are usually shunted to the side lines at international trade meetings or dismissed as out of touch in corporate public relations campaigns. And from the point of view of many of these firms, the logic of all this is plain: given their significant investments in research laboratories or in production facilities, the agricultural TNCs no doubt believe they deserve to reap what they sow and, as much as possible, without interference from the state (whether it is a democratic one or not). Concentrating decision-making power just seems to go with the terrain here: closing off options, evading regulation, keeping outsiders (that is, non- or small-scale shareholders) out of the picture is justified, from the point of view of these large TNCs, given the scale of the investments at stake. In addition, they might argue that many of the largest TNCs in other sectors of the capitalist economy – in manufacturing, say, or in financial services – are able to concentrate power and exert influence over democratic governments, so why should this not be the same in sectors focused on food production? 'What's fair for the car manufacturers should be fair for the producers of agricultural seeds; isn't this the nature of business?'

Against this line of argument is the claim that food should not be treated just like other commodities. It is different from a pair of jeans or the latest gadget because we all need to eat. The sorts of rules that might be ethically viable in other sectors of capitalism cannot apply to the life of food. Regarding the life of food and the role of the capitalist enterprises shaping it, we need a level of oversight, regulation, and inclusion that might not be quite so urgently needed in other sectors. The point of focusing on the degree to which decision-making power is concentrated in the boardrooms of TNCs is precisely to highlight the perils of excluding other interested and capable parties. We know mistakes have been made, that decisions about product innovations have led to adverse environmental effects or dangers to human health. The risk is that such mistakes will be repeated without greater oversight, regulation, or inclusion.

For Iris Young (1990), those who lack authority, those who experience the effects of decisions but have no control over the making of them, suffer *powerlessness*. This is not to say they lack power over all aspects of their lives; that they have no power or no potential to practise power. Consider here that power has various 'guises' (Allen, 2003): that one might not be able to shape certain decisions and therefore suffer powerlessness in that sphere of life, but that power also entails manipulation, say, or seduction, not to mention authority that might exist over other decisions. In other words, someone can suffer powerlessness in one sphere of life but retain power in many others. Nor should we imagine that an experience of powerlessness necessarily means hopelessness: the frustration of not shaping decisions is one thing, but for many people who struggle against oppressive structures or processes there will be belief that things can and might change.

I think we can justifiably argue that powerlessness in this first stage in the life of food is experienced by billions of people, given that only a small number of people – those in the boardrooms of corporations such as Monsanto or in legislatures such as the US Congress – are making major decisions about the sort of food we are all going to eat. Obviously, however, for many of the billions of people affected by the concentration of power in corporate boardrooms, their powerlessness won't matter much. Put simply, we are not all looking to shape decisions about the future of agricultural seeds or chemicals (even if we suggest that everyone *should* take an interest in what is taking place, which is hard to push too far given the various constraints under which people operate – not all of us have the time or the expertise to understand what is going on). But then again – and this is the crucial point – given the centrality of food to our lives as individuals and to human life more generally, there are many people who *are* interested *and* capable enough to shape decision-making about these issues, but yet they are powerless to intervene because the key decisions are made in private or in consultative procedures that do not fully include interested parties.

Consider here the case of Lee Kyung Hae, a peasant and farmer who committed suicide to use his body as a site of protest in 2003 outside a WTO ministerial meeting in Cancun, Mexico. Lee's story is well-known among those who ally themselves in some way with a large transnational agrarian movement called La Vía Campesina. Lee was alongside others in Vía Campesina protesting against the WTO and its drive to see tariffs and other so-called barriers to trade removed throughout the world and, as I have discussed earlier, especially in 'developing' countries. Lee and those other protesters were not part of the negotiations. Ministers were, strictly speaking, in Cancun to represent their citizens, but the fears and worries of many of their poorest citizens living in rural areas never seem to be their main concern. Rather, the WTO has pursued a world where food can be exported and imported without enough regard for how its movement will impact on those who have produced on the land for generations. Lee was one of many Koreans affected by new cattle imports from Australia. He borrowed, was unable to repay the loans, and then lost his farm (see Patel, 2007: 43–47). He was in Cancun alongside many others like him to protest against the uneven playing field that was actively generated by WTO rules.

Of course, sitting here with no knowledge of Lee's feelings at that time, I can hardly claim with certainty that he experienced a feeling of powerlessness; but his action was certainly an ultimate protest that reflected his inability to do anything about how the WTO and all of the various government ministers attending the Cancun meeting were largely disregarding the lives of peasant farmers like him. He was shut out. From the perspective of the ministers and lawyers and the corporate sector the WTO exists to support, Lee was unimportant; almost irrelevant. This is exactly what Iris Young meant by calling attention to powerlessness as a form of oppression. In highlighting Lee, moreover, I am thinking of all the other small-scale farmers or peasants who would like to shape the way the food system operates. Many of them are simply powerless to alter a system which promotes intensive agricultural practices and places their livelihoods at risk.

Furthermore, I think we need to dwell on the powerlessness experienced by all of the various but critical experts on ecology or nutrition, who might like to influence policy but cannot manage to get their views or opinions to count. A case in point is Vandana Shiva, the formidable critic of 'growth fetishism' and the use of GM technology in Indian agriculture, who consistently defends the rights of 'the poor and powerless women, farmers, tribals and the displaced rural communities who become urban slum dwellers' (Shiva, 2011) against the state and corporate sector. Her calls for change fall on deaf ears in the Indian state, despite her obvious expertise. For critical experts such as Shiva and for those immediately affected by the food system and by the decisions made upstream of the farm, their powerlessness has to be seen as deeply oppressive. The frustration of wanting to shape processes – of knowing your knowledge or perspective could be put to use but realizing that you cannot actually do anything – is deeply felt by many, not just with respect to seeds or agricultural chemicals. A feeling of insignificance, of powerlessness, hurts.

1.6 Resistance: imagining alternatives

As I noted earlier, the experience of powerlessness in one sphere of life does not mean an individual is powerless in all other spheres. Vandana Shiva might not be able to influence the Indian state as much as she wants, but she is influential in other spheres. Likewise, although Lee Kyung Hae suffered powerlessness with regard to the WTO, he had the capacity – the power – to make the ultimate statement of taking his own life in protest. Powerlessness might pertain in one sphere; but in numerous other spheres there might be hope, belief, and the capacity to act and do something. In this regard, what I refer to as 'imagining alternatives' is a crucial form of resistance against the experience of powerlessness. This is about the capacity to analyse and understand how powerlessness is generated and then the shift to imagining new structures or processes that might undo some of the harm.

One way we can see this upstream of the farm is with respect to contemporary debates about the development of knowledge over seeds. Now, to be clear and to be fair, we have to note here that knowledge about seeds today is *not* only a product of corporate investment. Much of the research and innovation (and then the eventual production) *does* occur thanks to the investment of these sorts of firms. But also worth highlighting are other 'enterprises' moving resources, commodities, information, and money to survive and possibly expand. Definitely the most numerous, but perhaps under-recognized, enterprise is the peasant household. It is, after all, the combined efforts of millions of peasant households that has allowed humans to select the most appropriate sorts of seeds to plant each year. Still today, for all of the talk in books like this about TNCs and their ilk, 80 per cent of the developing world's food comes from these sorts of small-scale agricultural households (International Fund for Agricultural Development, 2013). In addition, much of the work upstream of the production of food is performed by state and public sector bodies, such as regional or national agricultural research agencies, publicly funded university research, or international collaborations such as the Consultative Group on International Agricultural Research (CGIAR) group of research institutions (e.g. see Kloppenburg, 2004; Clapp, 2012: 34–40). The state – the set of agencies and institutions that spend public money – is clearly involved in the development of seeds, new chemicals, and new innovative ideas and practices related to food. Research institutions such as those under the CGIAR umbrella do produce major insights, not least about seeds, and certainly many of the regional and national research agencies matter. Thus, despite what capitalist suppliers of seeds and other farm inputs might say, it remains conceivable that agricultural sectors across the world could manage without the involvement of capitalist firms. As Vía Campesina and other advocates of agroecological approaches argue (see Chapter 2), agricultural dependence on the world of commercial inputs makes farmers vulnerable, and places the sustainability of the overall food production system in doubt (e.g. see Desmarais, 2007; Wittman *et al.*, 2010).

Still, profit-oriented transnational corporations *have* come to occupy highly prominent positions upstream of the farm. They look to create goods and services that increase their market share and deliver new sources of profit. Large, powerful firms are profiting from the human necessity of food production. When a new wave of innovation occurs, such as the development of GMOs, the end product might boost food production, but the real purpose of that development, from the point of view of the firms owning the intellectual property rights, is to generate future profits. And this is precisely what so many critics of the food system are alarmed about. What many people today find objectionable is that firms are so overwhelmingly in a position to conduct the sort of research that can lead to these innovations. The question many are now asking is: where is the state, the democratic sector, the oversight and regulation that can ensure these firms balance the needs of people and their environments with the profit motive? And, for many of these critics, the answer needs to come from bottom-up solutions: that is, grass-roots efforts that draw on networked and individual actions.

In this regard, one important form of resistance today is focused on finding and developing alternatives to corporate dominance of seeds. As noted earlier in this chapter, there is an enormous and growing market for seeds which is fundamentally shaped by firms such as Monsanto. In opposition to this market and against the powerlessness felt by many of those affected by the decisions of the leading firms, a diverse and committed movement has emerged, globally and locally, to bolster and create seed exchanges, seed banks, and other mechanisms to enable grass-roots and democratic control over seeds. In other words, many of those who want to shape the structures and processes via which seeds are distributed and used today but are unable to shape what Monsanto, say, or Pioneer actually do – that is, many of those who might say they suffer powerlessness – are involved in imagining and creating alternatives. There may be powerlessness in one sphere of life, but a strong sense of agency and scope to act in others.

The idea of seed exchanges has a long lineage globally, although they are not always known as such. In Malawi, for example, where people grow 'maize as their staple crop and groundnuts and other crops for sale and consumption' (Bezner Kerr, 2010: 134), seed is 'used as a gift or a type of currency in exchange for labour [and] women give maize and groundnut seed to their daughters-in-law upon marriage [and] instruct [them] about how to store and select seed for future harvest' (Bezner Kerr, 2010: 138–139). Likewise, in Northern Thailand, Karen communities 'jealously conserve native varieties of rice, grains, vegetables and herbs via rotational cultivation, and the Assembly of the Poor uses networks like the Alternative Agricultural Network to collect rice varieties, use agroecology methods and promote household consumption' (McMichael, 2010: 179). New moves exist as well, such as the effort by development NGO workers in Southern Chile to establish community seed banks in an area 'where more than 120 traditional [potato] varieties are grown year after year and are subjected to selection and seed enhancement' (Altieri, 2010: 128). And then in Norway, there is the Svalbard Global Seed Vault, which looks to secure access to crops for future generations by protecting millions of seeds at a temperate of –18 °C (Goldenberg, 2015) (see Figure 1.2).

All of these efforts are worthy, but for some scholars, perhaps most notably Jack Kloppenburg (e.g. see Kloppenburg, 2010a, 2010b), they do not go far enough, given the power and the drive of seed corporations. Rather, Kloppenburg suggests that another, more ambitious and certainly creative set of actions should involve developing a general public licence for plant germplasm (GPLG). The idea of a general public licence comes from the world of software production. Although firms such as Microsoft or Apple are central players in the development of software that they patent and can then sell, there is also an extensive world of software developers who offer their skills for free to develop what is known as 'open source'

Figure 1.2 Svalbard Global Seed Vault, Norway (source: dinozzaver/Shutterstock).

software: languages, protocols, and programs that are freely distributed and 'open' in so far as anyone with expertise can see and then change and develop the background data and code. Such software is not copyrighted as is, say, Microsoft Windows, but is rather given an open general public licence protected under a 'creative commons'. In a similar fashion, Kloppenburg's suggestion is to develop the same principles for distributed peer production regarding seed germplasm, much as movements for Open Source Biology have outlined. Rather than seed innovations becoming the property of private firms, seed scientists could release their findings in an open-source format in such a way that innovations can be shared for the general public good. As Kloppenburg notes, then, actions against corporate control over the food system, such as protests, need to be 'complemented by creative actions that are not just reactions to corporate/neoliberal conditions but which are offensive, affirmative, positive, pro-active undertakings designed to establish and maintain alternative, (relatively) autonomous spaces' (Kloppenburg, 2010a: 165). This is about imagining alternatives, which in many ways is one of the first steps to take in trying to stand up against an oppressive structure or system. As seed banks and exchanges and Kloppenburg's suggestions demonstrate, there is a bit of a fight-back against corporate control over seeds. And that there is some resistance should give us some hope amidst a bleak context in which corporations – and the material interests of their shareholders – seem to trump all other concerns, as we saw earlier in the case study of Monsanto. As I think all of this demonstrates, then, if there is one lesson to take from a consideration of this stage in the life of food, it is this: during conception and gestation – as indeed is the case all the way to consumption and digestion – the life of food is thoroughly contested and debated.

What else matters here?

CGIAR and its research institutes across the world matter when it comes to agricultural research. Can you find out more about what they do and what challenges they face today?

In the country where you live, there is bound to be some sort of government-funded support for farmers. Why not use newspaper archives to track recent developments: the big debates, such as over funding or policy; or the main controversies or stories developing in the last few years?

As I write this, there are serious debates about trade deals not unlike those that led to the WTO. What do you know about these deals? What are farmers, agribusinesses, or civil society organizations saying about them?

You decide

- *Should* food be treated just like all other commodities today? What might be some of the arguments for and against such a stance?
- How *should* we ensure there is sufficient investment in seeds or other agricultural inputs? Is reliance on the private sector necessarily problematic?
- Is democratic oversight of the global food system possible? More broadly, is democratic oversight of capitalist enterprises as a whole realistic?

Suggested reading

Chapters III, IV, and V of the 2009 World Investment Report deal in detail with many of the implications of agricultural TNCs and their role in the food economy.

I cannot recommend highly enough that you read *Global Food Economy* by Tony Weis (2007), but especially chapter 4, which so excellently reviews the emergence of the WTO. Also useful here is Haroon Akram-Lodhi's (2013) *Hungry for Change: Farmers, Food Justice, and the Agrarian Question.*

A groundbreaking and insightful take on the world upstream of the farm is Judith Carney's (2001) *Black Rice: The African Origins of Rice Cultivation in the Americas*, which examines the fundamental role of enslaved West Africans in developing the knowledge via which rice plantations were established in the Americas.

Notes

1 To a large extent, this is how China still does things: it sets limits on some imports and tariffs on others, all with a view to protecting its domestic agricultural sector (*Economist*, 2015). Any government could do the same, at least in principle, if not in practice in the contemporary period.
2 Many of these sorts of reforms came in the form of so-called structural adjustment programmes, which I discuss in more detail in Chapter 2.
3 For the sake of comparison, note here that the combined sales of the top 25 non-financial TNCs was $3.7 trillion; Shell, the third largest TNC in 2007, on its own had sales of $355 billion (UNCTAD, 2009: 255).
4 Worth noting as an aside here, as the recent book *The Informant* (Eichenwald, 2009) reviews, is that executives from ADM were recently alleged to have been heavily involved in setting up price-fixing meetings with competing firms in Japan and Korea. By agreeing to limit production and fix prices, Eichenwald notes, these firms tried to generate unusually high profit rates by forcing farmers to pay over the odds. ADM and all of the large firms like it have these opportunities open to them – even if they do not take advantage of them – in part because they dominate their markets.

References

Akram-Lodhi, A.H. (2013) *Hungry for Change: Farmers, Food Justice, and the Agrarian Question*. Halifax: Fernwood.

Allen, J. (2003) *Lost Geographies of Power*. Oxford: Blackwell.

Altieri, M.A. (2010) Scaling Up Agroecological Approaches for Food Sovereignty in Latin America. In: Wittman, H., Desmarais, A.A., and Wiebe, N. (eds), *Food Sovereignty: Reconnecting Food, Nature & Community*. Oxford: Pambazuka, pp. 120–133.

BASF AG. (2014) *BASF Report 2014: Economic, Environmental and Social Performance* [online]. Available at: http://bericht.basf.com/2014/en (accessed 22 January 2016).

Bayer AG. (2014) *Annual Report 2014* [online]. Available at: www.annualreport2014.bayer.com/en/bayer-annual-report-2014.pdfx (accessed 22 January 2016).

Bezner Kerr, R. (2010) Unearthing the Cultural & Material Struggles over Seed in Malawi. In: Wittman, H., Desmarais, A.A., and Wiebe, N. (eds), *Food Sovereignty: Reconnecting Food, Nature & Community*. Oxford: Pambazuka, pp. 134–151.

Brown, W. (2015) *Undoing the Demos*. Brooklyn, NY: Zone.

Cargill. (2015) *Five-year Financial Summary* [online]. Available at: www.cargill.com/company/financial/five-year (accessed 22 January 2016).

Carney, J. (2001) *Black Rice: The African Origins of Rice Cultivation in the Americas*. London: Harvard University Press.

Carolan, M. (2011) *The Real Cost of Cheap Food*. Abingdon: Earthscan.

Carrington, D. (2014) Insecticides Put World Food Supplies at Risk [online]. *Guardian*. 24 June 2014. Available at: www.theguardian.com/environment/2014/jun/24/insecticides-world-food-supplies-risk (accessed 22 January 2016).

Clapp, J. (2012) *Food*. Cambridge: Polity.

Desmarais, A.A. (2007) *La Vía Campesina: Globalization and the Power of Peasants*. London: Pluto.

Dow Chemical. (2014) *Annual Report 2014* [online]. Available at: www.dow.com/~/media/DowCom/Corporate/PDF/investor-relations/2014-Dow-Annual-Report-with-10K.ashx?la=en-US (accessed 22 January 2016).

Drew, J. and Joseph, J. (2012) *The Story of the Fly: And How It Could Save the World*. Vlaeberg: Cheviot Publishing.

Drew, J. and Lorimer, D. (2011) *The Protein Crunch*. Noordhoek: Print Matters Planet.

Du Pont. (2014) *Data Report 2014* [online]. Available at: http://investors.dupont.com/files/doc_financials/2014/Databook-2014-FINAL.pdf (accessed 22 January 2016).

The Economist. (2015) Bitter Harvest [online]. *The Economist*. 16 May 2015. Available at: www.economist.com/news/china/21651276-drive-self-sufficiency-food-comes-growing-cost-bitter-harvest (accessed 22 January 2016).

Eichenwald, K. (2009) *The Informant*. London: Portobello Books.

Ferrara, J. (1998) Revolving Doors: Monsanto and the Regulators. *The Ecologist*. Available at: www.psrast.org/ecologmons.htm (accessed 22 January 2016).

Food and Agriculture Organization. (1999) *Sources of Funds for Agricultural Lending*. Rome: Food and Agriculture Organization of the United Nations. Available at: ftp://ftp.fao.org/docrep/fao/012/ak920e/ak920e00.pdf (accessed 22 January 2016).

Friedmann, H. (1993) The Political Economy of Food: A Global Crisis. *New Left Review*, 197 (January–February), 29–57.

George, S. (2015) *Shadow Sovereigns: How Global Corporations are Seizing Power*. Cambridge: Polity.

Goldenberg, S. (2015) The Doomsday Vault: The Seeds That Could Save a Post-Apocalyptic World [online]. *Guardian*. 20 May 2015. Available at: www.theguardian.com/science/2015/may/20/the-doomsday-vault-seeds-save-post-apocalyptic-world (accessed 22 January 2016).

Harvey, D. (2005) *A Brief History of Neoliberalism*. Oxford: Oxford University Press.

International Fund for Agricultural Development. (2013) *Smallholders, Food Security, and the Environment* [online]. London: International Fund for Agricultural Development. Available at: www.unep.org/pdf/SmallholderReport_WEB.pdf (accessed 22 January 2016).

Klein, N. (2007) *The Shock Doctrine*. London: Penguin.

Kloppenburg, J.R. (2004) *First the Seed: The Political Economy of Plant Biotechnology* (2nd edn). Madison, WI: University of Wisconsin Press.

Kloppenburg, J. (2010a) Seed Sovereignty: The Promise of Open Source Biology. In: Wittman, H., Desmarais, A.A., and Wiebe, N. (eds), *Food Sovereignty: Reconnecting Food, Nature & Community*. Oxford: Pambazuka, pp. 152–167.

Kloppenburg, J. (2010b) Impeding Dispossession, Enabling Repossession: Biological Open Source and the Recovery of Seed Sovereignty. *Journal of Agrarian Change*, 10(3), 367–388.

Liptak, A. (2013) Supreme Court Supports Monsanto in Seed-Replication Case [online]. *New York Times*. 14 May 2013. Available at: www.nytimes.com/2013/05/14/business/monsanto-victorious-in-genetic-seed-case.html (accessed 22 January 2016).

Macilwain, C. (2015) Rejection of GM Crops is Not a Failure for Science [online]. *Nature*. 2 September. Available at: www.nature.com/news/rejection-of-gm-crops-is-not-a-failure-for-science-1.18271 (accessed 22 January 2016).

Marsden, T. (2013) Contemporary Food Systems: Managing the Capitalist Conundrum of Food Security and Sustainability. In: Murcott, A., Belasco, W., and Jackson, P. (eds), *The Handbook of Food Research*. London: Bloomsbury, pp. 135–147.

Martin, S.J. and Clapp, J. (2015) Finance for Agriculture or Agriculture for Finance? *Journal of Agrarian Change*, 15(4), 549–559.

McMichael, P. (2005) Global Development and the Corporate Food Regime. *New Directions in the Sociology of Global Development: Research in Rural Sociology and Development*, 11, 269–303.

McMichael, P. (2006) Peasant Prospects in the Neoliberal Age. *New Political Economy*, 11(3), 407–418.

McMichael, P. (2010) Food Sovereignty in Movement: Addressing the Triple Crisis. In: Wittman, H., Desmarais, A.A., and Wiebe, N. (eds), *Food Sovereignty: Reconnecting Food, Nature & Community*. Oxford: Pambazuka, pp. 168–185.

Monsanto (n.d.) Round Up Ready System [online]. Available at: http://test.monsanto.com/weedmanagement/pages/roundup-ready-system.aspx (accessed 13 April 2016).

Norton, R.D. (2004) *Agricultural Development Policy*. Chichester: John Wiley.

OECD (Organisation for Economic Co-operation and Development). (2015) *Agricultural Policy Monitoring and Evaluation Report* [online]. Paris: Organisation for Economic Co-operation and Development. Available at: www.oecd.org/tad/agricultural-policies/monitoring-evaluation-2015-highlights-july-2015.pdf (accessed 22 January 2016).

Paarlberg, R. (2010) *Food Politics: What Everyone Needs to Know*. Oxford: Oxford University Press.

Paarlberg, R. (2013) The World Needs Genetically Modified Foods [online]. *Wall Street Journal*. 14 April 2013. Available at: www.wsj.com/articles/SB100014241 27887324105204578380872639718046 (accessed 22 January 2016).

Pardey, P.G., Chan-Kang, C., Beddow, J.M., and Dehmer, S. (2014) Long-run and Global R&D Funding Trajectories: The U.S. Farm Bill in a Changing Context. Agricultural and Applied Economics Association, 2015 Allied Social Sciences Association (ASSA) Annual Meeting, 3–5 January 2015, Boston, MA. Available at: http://purl.umn.edu/189694 (accessed 22 January 2016).

Patel, R. (2007) *Stuffed and Starved: Markets, Power and the Hidden Battle for the World Food System*. London: Portobello.

Reuters. (2015) Global Dependence on Food Imports Leaves Countries Vulnerable [online]. *Reuters*. 13 March 2015. Available at: http://uk.reuters.com/article/2015/03/13/uk-food-trade-idUKKBN0M926820150313 (accessed 22 January 2016).

Scoones, I. (2008) Mobilizing Against GM Crops in India, South Africa and Brazil. In: Borras, S.M. Jr., Edelman, M., and Kay, C. (eds), *Transnational Agrarian Movements Confronting Globalization*. Oxford: Wiley-Blackwell, pp. 147–175.

Shiva, V. (2011) Who Pollutes: The Rich and Powerful or Poor and Powerless? [online]. *Z Comm*. 22 July 2011. Available at: https://zcomm.org/zcommentary/who-pollutes-the-rich-and-powerful-or-poor-and-powerless-by-vandana-shiva (accessed 22 January 2016).

UNCTAD (United Nations Conference on Trade and Development). (2009) *World Investment Report: Transnational Corporations, Agricultural Production and Development*. New York, NY: United Nations.

Weis, T. (2007) *The Global Food Economy: The Battle for the Future of Farming*. London: Zed Books.

Wittman, H., Desmarais, A.A., and Wiebe, N. (2010) *Food Sovereignty: Reconnecting Food, Nature & Community*. Oxford: Pambazuka.

World Bank. (2007) *World Development Report 2008: Agriculture for Development*. Washington, DC: World Bank.

World Trade Organization. (2014) *International Trade Statistics 2014* [online]. Geneva: World Trade Organization. Available at: www.wto.org/english/res_e/statis_e/its2014_e/its2014_e.pdf (accessed 22 January 2016).

Young, I.M. (1990) *Justice and the Politics of Difference*. Princeton, NJ: Princeton University Press.

2 Agricultural foodscapes

2.1 Introduction

What I tried to do in Chapter 1 was develop a sense for how food production is shaped by action that unfolds before anything happens on the farm. I spent time discussing the governance of agriculture and then turned to the role played by broker and supplier firms in developing services and products that connect with the farm and seek to profit from the act of producing food. That upstream world is important; paying attention to it – and to the sorts of foodscapes we find there – reveals interesting features about how governments, capitalist firms, and farmers have tried to answer the proletarian food question under the rules and practices of the corporate food regime.

In this chapter I want to proceed from that upstream world and, in effect, move *down*stream to the spaces and places where food production occurs. I think our attention in this second stage in the life of food needs to be on 'northern agriculture' – this relatively new way of using the land which tends to rely on government subsidies, constantly uses a wide range of purchased inputs, and increasingly needs a worldwide market for sales. I therefore discuss how northern agriculture has emerged, and highlight some prominent features of the foodscapes it produces. I then try to dwell upon what northern agriculture means for all of us on this planet and, in doing so, I especially consider how it relates to the 2.5 billion or so people living in smallholder or agricultural households across the so-called 'global south' (World Bank, 2007: 3). More specifically, I identify a set of logics driving the geographical expansion of northern agricultural practices; discuss how they are playing out today; and note some of the most alarming consequences.

The rest of the chapter is organized as follows. In Section 2.2 I provide some background data on food production that help us identify and locate where northern agriculture exists. I then discuss how northern agriculture has taken shape and zoom in on some of its most striking aspects. In Section 2.3 I consider what northern agriculture and the corporate food regime means, and in particular what its geographical expansion might lead to. The case study in Section 2.4 on the expansion of soybean production in

Paraguay examines an indicative contemporary development emerging from the corporate food regime's underlying logics. Section 2.5, which focuses on how this stage in the life of food is oppressive, uses Young's (1990) concept of violence to shed light on some of the most problematic features emerging from this stage in the life of food. Finally, Section 2.6 asks how this overall picture is resisted, focusing on the use of 'occupy' as a resistance strategy that can challenge the logics that create an unjust and imbalanced global foodscape.

2.2 The foodscapes of northern agriculture

It might help if we begin by briefly reviewing some background data about food production. According to the Rome-based and United Nations-funded Food and Agriculture Organization, there are 4.7 billion hectares of agricultural land in use today, distributed among 569 million agricultural holdings (Table 2.1). Notably, only 3.84 per cent of these holdings – we might also call them 'farms', 'units', or 'plots' – are in high-income countries, yet they make up 26.1 per cent of the world's total agricultural land. This indicates that agricultural holdings tend to be larger in high- as opposed to low-, lower-middle-, or upper-middle-income countries. The data in Table 2.2 further demonstrate that 82 per cent of the land area in high-income countries is taken up by 9 per cent of the holdings, that is, those with more than 50 hectares;[1] and a similar pattern exists in upper-middle-income countries, too. Meanwhile, in low-income and lower-middle-income countries respectively, 73 and 57 per cent of the agricultural area is taken up by holdings smaller than five hectares.

Table 2.1 Population, agricultural holdings, and agricultural area in low- to high-income countries

	2012 population (millions)	*Percentage of the total*	*Holdings (thousands)*	*Percentage of the total*	*2011 agricultural area (thousand ha)*	*Percentage of the total*
Low-income	846	12.0	71 522	12.6	619 851	13.0
Lower-middle-income	2507	35.6	208 148	36.5	837 233	17.6
Upper-middle-income	2391	33.9	268 035	47.1	2 063 966	43.3
High-income	1302	18.5	21 867	3.8	1 246 991	26.1
World	7046	100.0	569 600	100.0	4 768 186	100.0

Source: FAO (2014, table A1); World Bank (2014b, table 1).

Table 2.2 Shares (per cent) of agricultural holdings by size class in low- to high-income countries

		<1 ha	*1–5 ha*	*5–20 ha*	*20–50 ha*	*>50 ha*
Low-income	Holdings	63	33	4	0	0
	Area	20	53	25	1	2
Lower-middle-income	Holdings	62	33	5	0	0
	Area	15	42	24	8	11
Upper-middle-income	Holdings	27	42	21	6	5
	Area	0	4	7	7	81
High-income	Holdings	34	33	16	7	9
	Area	1	3	6	8	82
World	Holdings	72	22	4	1	1
	Area	8	11	10	7	65

Source: FAO (2014, table A2).

If one difference between high-income and poorer countries is farm sizes, there are numerous others. One particularly noticeable difference is the high labour productivity per worker in high-income countries (Table 2.3), which is 55 times higher than in low-income and 18 times higher than in upper-middle-income countries. As Table 2.4 demonstrates, moreover, the figure of $27 112 for labour productivity in high-income countries is relatively low when compared with productivity rates in some countries, with Belgium and then, more importantly, the US – which I think we need to view as the heartland of northern agriculture – really leading the way. Finally, as indicated in Table 2.5, agriculture in these high-income, highly productive countries has more tractors and tends to use more fertilizer than elsewhere: thus, as we will see, variation in the level of input-use – in the availability of technology and the money to afford it – goes a long way to explain the differences in productivity rates.

The 'northern agriculture' I am referring to, then, is the type of agriculture we find concentrated in high-income countries, and scattered more loosely throughout lower-income countries. It is a way of using the land that generates massive yields – for example, between 2008 and 2012 the cereal yield per hectare in the US averaged 6716 kg (World Bank, 2014a), whereas yields in low-income countries in 2013 averaged 1581 kg. But the high yields of northern agricultural practices are not the only noteworthy feature here. Rather, and as I discussed in Chapter 1, this is a government-backed, policy- and market-driven agriculture: something that has been created via purposeful action, not least with regard to subsidies, supports, and then negotiations at a global scale that sought to cement the advantages and powers of the rich world's agricultural system. This is a system that works for the supplier firms, as well as for many of the buyers of agricultural produce in the global north.

Table 2.3 Average agricultural labour productivity in low- to high-income countries, 1961–2012

	Average annual level (constant 2004–2006 international dollars)				
	1961–1971	*1971–1981*	*1981–1991*	*1991–2001*	*2001–2012*
Low-income	405	412	416	419	490
Lower-middle-income	748	848	937	902	1 057
Upper-middle-income	527	609	720	1 003	1 454
High-income	5 556	8 627	12 211	18 095	27 112
World	943	1 059	1 141	1 261	1 535

Source: FAO (2014, table A3).

Table 2.4 Average annual agricultural labour productivity in selected high-income countries, 1961–1971 and 2001–2012

	Average annual level (constant 2004–2006 international dollars)	
	1961–1971	*2001–2012*
Australia	25 721	51 981
Belgium	...	81 004
Canada	13 527	68 306
Denmark	13 504	69 608
France	8 651	57 626
Germany	6 538	41 180
Netherlands	17 006	53 204
New Zealand	37 078	53 997
UK	14 465	32 257
US	23 145	74 723

Source: FAO (2014, table A3).

Table 2.5 Use of fertilizer and farm machinery in low- to high-income countries

	Fertilizer consumption (kilograms per hectare of arable land, 2005–2007)	*Tractors per 100 km² of arable land*	
		1990–1992	*2005–2007*
Low-income	35.0	33.5	33.7
Lower-middle-income	155.1	72.3	140.7
Upper-middle-income	70.5	252.6	175.2
High-income	143.8	360.2	380.7
World	117.7	189.6	198.7

Source: World Bank (2010, table 3.2).

Notably, the structure of northern agriculture – the particular arrangement – is not necessarily something that is warmly embraced by farmers who practise it. There is, in a sense, and as Weis (2007) reviews in detail, a battle over the future of farming, which pits the corporate sector – and the basis of a food regime oriented towards corporate profits – against many smaller-scale farmers. There is serious discontent, not to mention exhaustion and frustration, about the way agriculture has developed over the last few decades. Although, as I also noted in Chapter 1, part of the problem is undoubtedly that subsidy regimes in the US and Europe disproportionately benefit large-scale operations, which generates a sense of injustice among smaller-scale farmers; a more enduring problem is about how agriculture has been captured and squeezed by firms operating up- and downstream of the farm.

One way to come to terms with all this is to consider the concept of input-dependence as it applies to agriculture. Farming always needs inputs, of course. Crop farmers need seeds and often fertilizer. Livestock farmers need feed. And today – in rich and poor countries – there are still food producers who use their own seeds, manure, and grow the crops fed to livestock. Many organic farmers, for example, try to create a self-renewing system with functional diversity, multi-cropping, pasturing, and green manures (following Weis, 2007: 29; see also Guthman, 2004). In other words, organic farmers look to produce food via a modest, intimate, and sustainable relationship with the land. However, and in stark contrast, waves of scientific and technological developments – often far from the farm, sometimes without food production in mind – have introduced radical new ideas that technological fixes can consistently boost farm output.

Consider seeds. Up until the 1950s, agricultural producers mostly used their own or locally available seeds. But public and private investments in seed technology – part of the so-called Green Revolution – led to new 'miracle' hybrid high-yield seeds that were cross-bred, researched, tested, studied, and then in many cases patented (Kloppenburg, 2004; Weis, 2007: 70–72). Some of these hybrid, high-yield seeds better handled climate variability than 'traditional' varieties, or they were selected to suit specific artificial conditions, such as the application of irrigated water or particular chemicals. They appealed to many farmers, especially when they led to higher yields and, maybe for a period, higher incomes. Today, of course, seed technology has moved up a gear to look not just at breeding, but rather at the manipulation of the microscopic genetics of plants. These new genetically modified organisms (GMOs) are a source of controversy, in part because patent protections make seeds a non-renewable resource that must be purchased year after year; and in part, too, because science is unsure whether GMOs will spill over into the environment and have unforeseen negative consequences for other species. Alongside these dimensions of a debate about seeds is the fact that just a few firms – Pioneer Hi-Bred, Shell, Monsanto, Dow, and Dupont – increasingly dominate the commercial seed market (see Clapp, 2012: 38). Indeed, according to McMichael (2010: 177–179), this process of seed commercialization is the 'principal legacy' of the Green Revolution, one which shapes the global food economy in fundamental and problematic ways: not least by giving the seed supplier firms financial power to shape how agriculture is regulated and governed. Notably, of US crop farmers' production expenditures in 2014, $19 billion (9.39 per cent of the total) went on seeds (USDA, 2014) – this is a major expense for farmers and a large market that needs to be supplied.

After seeds, a second technological fix farmers using northern agricultural practices rely upon is synthetic fertilizer. Whereas farmers once relied on creating their own manure or buying from a merchant – such as those who brought human excrement from urban areas to the countryside (e.g. in Korea, see Henry, 2005) – the logic of northern agriculture, which emphasizes output growth and yield, dictates that crop farmers buy the latest and greatest product on the market. Since the beginning of the twentieth century this has been about non-organic fertilizers – based on nitrogen, phosphate or potash – produced by firms such as BASF (Weis, 2007: 55–56). Their use has boosted agricultural production, arguably even making twentieth-century human population growth possible (Sage, 2012: 45–46). On application they provide an immediate boost to output by replacing nutrients that are otherwise lost while crops grow. This enables farmers to abandon crop rotation, which requires planting clover or beans to fix nitrogen in the soil for a cereal crop that will be planted the following year (Sage, 2012: 45). As crop rotation is abandoned, farmers are able to increase the total amount of productive land in use in any one year and open up areas of land that might otherwise have not been so productive. In effect, this allows soil to be 'mined', i.e. used up, never to be replenished but compensated by the use of fertilizer (Weis, 2007: 54–59). It is, therefore, easy to understand why use of fertilizers has increased dramatically in the last 50 years or so: for example, whereas in 1961, 31.2 million tonnes of non-organic fertilizer were used globally, this figure had reached 141.6 million tonnes by 2005 (Millstone and Lang, 2008: 48–49), and there is every reason to expect their use to continue growing. In 2014, US crop farmers spent $23.2 billion on fertilizers (11.5 per cent of their total production expenditures) (USDA, 2014).

A similar story of technological development explains the growing use of a third input: agricultural chemicals. These chemicals – 'elixirs of death' (Carson, 1962: 31–49) such as herbicides, fungicides, and pesticides – attack insect or plant species that detract from farm productivity, and generally enable farmers to reduce labour inputs, for example that were once needed to remove weeds by hand or deal with insects on a daily basis (Millstone and Lang, 2008: 46–47; Sage, 2012: 46–47). Northern agriculture depends heavily on pesticides, with US crop farmers spending $14 billion on chemicals in 2014 (7 per cent of total expenditures) (USDA, 2014). But as Rachel Carson (1962: 25) explained about the chemical DDT (dichlorodiphenyltrichloroethane) in her landmark book *Silent Spring*, farmers must continue to use them, often on an increasing scale:

> The whole process of spraying seems caught up in an endless spiral. Since DDT was released for civilian use, a process of escalation has been going on in which ever more toxic materials must be found. This has happened because insects, in a triumphant vindication of Darwin's principle of the survival of the fittest, have evolved super races immune to the particular insecticide used, hence a deadlier one has always to be developed – and then a deadlier one than that. It has happened also because ... destructive insects often undergo a 'flareback', or resurgence, after spraying, in numbers greater than before. Thus the chemical war is never won, and all life is caught in its violent crossfire.

This 'endless spiral', in turn, produces dependence, which suits the producers of agricultural chemicals, and generates a highly toxic agricultural landscape and results in thousands of annual deaths from accidental (and in some cases, deliberate suicidal) ingestion (e.g. see Patel, 2007: 23–27)

A fourth input on which crop farmers in northern agriculture have become dependent is farm machinery, produced by firms such as John Deere from the US, with average annual sales of $33.8 billion between 2012 and 2014 (Deere, 2014), or Kubota from Japan with its 2014 sales of $10.4 billion (Kubota, 2014). Whether the combine-harvester or the water pump and sprinkler used in irrigation (Millstone and Lang, 2008: 34–35; Sage, 2012: 41–44), it is relatively easy to grasp how machinery can help boost output. Cutting through thick soils, moving fodder, gathering crops: all such activities can be done faster and with less labour or draught power. In replacing people and animals, tractors and combine-harvesters have enabled large plots of land to be ploughed and sowed quickly but with the necessity also of purchasing fuel. Such a process of 'motomechanization' has gathered pace in the last 50 years in particular, thereby allowing one worker to prepare around 200 hectares of land today, as opposed to just 30 hectares by the early 1950s (Sage, 2012: 42, citing Mayozer and Roudart, 2006). Notably, although the total number of tractors in use today has not increased much since the early 1990s, they have grown more powerful and sophisticated (Millstone and Lang, 2008: 34–35). And mechanization has extended to new areas of agriculture, such as dairy farms where technologies have enabled farmers to rapidly increase the number of cows each worker can milk (Sage, 2012: 41). Consequently, the number of people employed in agriculture has rapidly declined in the last few decades. For example, whereas about 20 per cent of the total workforce were involved in agriculture in the US in 1930 (Weis, 2007: 82), this figure has fallen to below 2 per cent (World Bank, 2014a). For some, no doubt, this is a progressive shift: reflecting a sense that progress is about society moving away from forms of menial, basic labour towards specialized, technology-intensive and high-skill work. Agriculture's declining importance in the labour market as a whole might be something to be celebrated (so long as such 'modern' types of work can be found, which is far from inevitable in today's economy). Even so, many farmers still need workers, sometimes drawn from the local area, but in other instances non-local and even immigrant labour, such as farm workers from Honduras, Guatemala, Mexico, Puerto Rico, and Jamaica who have migrated to look for work on farms across the US and Canada. Although some have been facilitated by government-sponsored farm worker programmes that provide working visas, many others endure horrendous conditions to cross borders and find work without documentation (e.g. on California, see Walker, 2004: 66–75).

So far I have considered crop farmers and their input-dependence. Livestock farmers encounter similar dynamics. Mechanization matters here because, as Weis (2007: 57) notes, it has 'allowed for the disarticulation of farm animals and crops'. Not only do farmers need fewer animals for draught power, but they no longer need to reserve land for pasture. In turn, this changing relationship creates scope for farmers to specialize in the production of a few crops, or to focus only on producing meat. In response to this change, the foodscapes of meat production have evolved in dramatic ways over the last few decades. An entirely new technological infrastructure has emerged: the so-called 'factory farm' – or, concentrated animal feed operations (CAFOs). These sites where chickens, pigs, and now even cows are produced stand out as the most industrial of all the spaces of food production associated with northern agriculture, not least because of the sheer number of animals produced and shipped off to be slaughtered: in 2014 in the US alone, for instance, 30.2 million cattle, 591 000 calves, 107 million hogs, and 8.6 billion chickens were slaughtered (USDA, 2015a, 2015b). Inside CAFOs, too, the scene is quasi-industrial: there are tubes circulating overhead to deliver feed and water; machines scooping up excrement; even floors that double as scales to measure how much chickens, say, have grown. Animals reared under these conditions live artificially short lives in close confinement at densities so high that disease spreads rapidly. Moreover, as Weis (2013: 141) notes:

> Environments of concrete and steel lead to extreme sensory deprivation and monotony. Animals are cut off from daily and seasonal rhythms and the ability to breathe fresh air, play, explore, or find food, sun, or shade, and are faced with frustrated co-inhabitants instead of families, social groups, and playmates. Painful musculoskeletal, cardiovascular, and respiratory disorders add to the misery in these crowded, noisy, reeking, and artificially lit spaces.

In response, meat producers look for technological fixes, such as antibiotics – produced by large transnational firms such as Zoetis – that offer animals some protection from the side-effects of living in close confinement in their own faeces. Yet, because some bacteria have become resistant to these drugs, demand for antibiotics has increased, with some CAFOs giving animals a constant intake, much to the concern of public health experts who argue that human health is adversely affected by exposure to antibiotics from the food chain (Weis, 2013: 138). Other drugs are needed, too, for example to deal with reproductive difficulties and stress-induced behavioural pathologies (Weis, 2013: 125). This market for livestock pharmaceuticals is large ($20 billion in 2004 (Weis, 2007: 72)) and growing, especially as meat consumption gathers pace and CAFOs struggle to contain infection and disease among their stock (Nierenberg, 2005).

Meat production in these sites also requires the supply of feed on an enormous scale, with livestock farmers in the US spending $62.1 billion on feed in 2014 (USDA, 2014). Today, indeed, as much as 70 per cent of all grains produced in the US are fed to livestock and 14 times as much land in the US is dedicated to producing hay, a feed crop, as is used to grow vegetables consumed by people (Weis, 2007: 42, 64). Of the US corn crop, around 80 per cent is fed to domestic and overseas livestock (Environmental Protection Agency, 2014). In other words, vast quantities of food that humans could eat – food that hungry people could eat – is used to feed animals. And part of the problem here is that the overall food system loses food as it is converted into meat: whether it is 3 kg of grain to produce 1 kg of chicken, or 6 kg to produce 1 kg of beef, the point is that producing meat is a massive waste of food.

A final feature here is that, whereas farmers might once have relied simply on their own grass or crops to feed livestock a relatively straightforward diet (Millstone and Lang, 2008: 38–39), they now purchase feed products on the market that contain drugs intended to boost fertility or improve immunity to infection and disease and even growth hormones intended to boost animal growth and bring them to market quicker. What the use of hormones speaks to is the overriding idea driving CAFOs: speed up turnover time, create products faster, and ensure that any money invested in the various inputs is returned at an ever-increasing rate, thereby also increasing the absolute amount of profit in a given period of time. It reflects a logic that we might expect to see when we look at how cars, smartphones, or clothes are produced, but not when we imagine the production of food and certainly not when animals are involved. Yet, these are the violent foodscapes created to get meat onto our plates today. Northern agriculture has emerged in part via a process in which farm animals have been 'progressively simplified into pure food commodities' (Weis, 2007: 57). This process of simplification has entailed the creation of nightmarish and complex sites that require a constant supply of inputs, the production of which entrenches the power of supplier firms upstream of the farm, as well as the overall corporate food regime overseeing food production today.

Overall, then, I think we need to view the emergence and growing dominance of northern agriculture as a radical and revolutionary change. It is a product of diverse policies, plans, and projects from government, as well as the actions of corporations and indeed farmers. What is often viewed now as 'conventional agriculture' is a way of farming that draws on technology to surpass limits; to expand output far beyond what was historically possible. This is a rationality that views technological fixes as the solution to problems that might emerge on the farm. Rather than viewing and using the land in a modest way, agricultural practices today increasingly assume that technology and science can imagine and deliver solutions that will continue to boost productivity. And in so far as many farmers, whether growing food crops or rearing animals, are operating at a scale and with an intensity we might expect to find in a factory, we therefore have to imagine northern agricultural systems as quasi-industrial. Just as a factory might need a constant supply of electricity, many farmers also require a steady supply of the inputs they use. Without the seeds, the fertilizer, or the antibiotics, the rates of output to which farmers have become accustomed would be unachievable. Farming the land has become intimately bound up with the need to purchase these inputs, almost as if, as Weis (2007) evocatively suggests, farmers are stuck on a *treadmill*: running to stay still but constantly using energy and spending money to do so.

Thus, when we look out at the agricultural foodscape today, we *might* still see something that resembles a rural idyll and have a sense that nature is at work: a sense that, here on the farm are humans and the natural world working together to create the food we all need. There is still some accuracy to that image, without question, and not just because there are still some farmers rearing cows on grass rather than commercial fodder, and other farmers opting to steer clear of relying on synthetic inputs; not all farmers have the same ideas about their place in society. But at the same time, the trend over the last 50–60 years has been for an industrial style of northern agriculture to become dominant, reflecting the rise of an input-dependent system that generates output and delivers a return on investment, while reproducing a broader and problematic corporate food regime.

The clear result is growing dependence on a wide range of inputs produced upstream of the farm, much of which is controlled by a relatively small number of transnational corporations. This trend towards greater input-dependence is absolutely ideal from the perspective of all the suppliers, many of them transnational firms with sales throughout the world. Given all this, it is clear that powerful transnational firms are deeply invested in seeing a continued expansion of northern agricultural practices and further consolidation of the corporate food regime on which it builds. Their shareholders demand consistent annual growth, both in revenues and in profits. Without growth, shareholders will be likely to stop investing in these firms and move on to buy other shares. Likewise, if rising revenues or profits are not delivered, shareholders will sell up, which in turn will lead to falling stock prices, which can then leave those TNCs vulnerable to a takeover from a rival firm. Food corporations are vulnerable, at least in this sense, because so much of their capacity to invest and innovate relies not just on sales but also on outside investors having faith in their future prospects. Consequently, all of the sorts of firms we have charted so far – those producing inputs such as fertilizer and tractors – need an expanding market: if not more farmers, then at least more sales from each of those farmers. It follows from all this that the logical outcome of the corporate food regime is for the big winners – the transnational corporations supplying farms, the traders, and then the supermarkets – to push for further expansion, including for more free trade agreements to give them access to new markets, as indeed the current negotiations over the Pacific and Atlantic free trade agreements demonstrate (see *New York Times*, 2015; *Reuters*, 2015).

2.2.1 Northern agriculture and the environment

Northern agriculture – with its massive cereal yields and animal slaughter on an immense scale – goes a long way towards answering the proletarian food question. It might be tempting to focus only on this aspect and, well, just leave it at that. But its emergence generates some significant environmental challenges, which deserve our attention. One of these is about the relationship between northern agriculture and climate change. Synthetic nitrogen-based fertilizers release nitrous oxide (N_2O) into the atmosphere, thereby forming tropospheric ozone, an extremely potent greenhouse gas (GHG). There has been significant growth in fertilizer use in the last few decades, especially in Asia, where use has jumped from 'an annual average of 6 kilograms per hectare in 1961–63 to 143 kilograms per hectare in 2000–02' (World Bank, 2007: 51). On the one hand, of course, this sort of growth rate has helped meet rising demand. But as Sage (2012: 117) notes, the problem is that 'crops may not take up even half the nitrogen made available through the application of fertilizer; the remainder is either leached through the soil creating pathways of nitrate contamination in ground and surface water, or is released as N_2O into the atmosphere'. In so far as northern agriculture relies heavily on fertilizers, food production based on its practices has further potential to generate climate change.

Then we need to bear in mind livestock production. Livestock generate methane emissions – a powerful GHG – due to enteric fermentation, which happens as microorganisms break down carbohydrates and allow animals to absorb them. The problem here is that livestock production has boomed in the last few decades. For example, as Weis (2007: 17) notes, 'while the human population has more than doubled since 1950, meat production grew nearly five-fold by volume' (Weis, 2007: 17). This jump in meat production satisfies growing consumer demand for meat – mostly beef, chicken, and pork. In 1960, US consumers ate an average of 85 kg of meat, but they now eat 106 kg, with chicken's share increasing from just 10 kg in 1960 to 43 kg today (Earth Policy Institute, 2014). Further meatification of diets, which seems likely given rising incomes in Asia in particular, will only add to the problem – with the only alternative here a radical re-thinking of meat consumption (Weis, 2013).

A final environmental question about northern agriculture is the extent to which it relies on using groundwater resources faster than they can be regenerated. The amount of irrigated land has doubled since 1950 (Millstone and Lang, 2008: 24), which has led some areas to run out of groundwater or struggle to manage conflicts over any water that is left over. In the US, the Ogallala aquifer, which lies beneath one of the heartlands of northern agriculture, has been heavily depleted by agricultural irrigation, with some parts of the aquifer emptied entirely (Weis, 2007: 59; Sage, 2012: 121–126). Meanwhile, food production in many other areas relies instead on river water, often water flowing from other regions or countries, which raises obvious dilemmas about how access to water should be managed and conflicts minimized. To the extent that key zones of northern agriculture rely on irrigation, they require an input whose sustainability is in question, which in turn raises numerous problems about how food production can overcome new challenges and avoid new crises.

In highlighting these environmental issues, I do not wish to suggest that some form of non-capitalist agriculture is necessarily going to have a better environmental record. Given environmental changes, such as widespread deforestation in Europe, long before capitalism emerged (e.g. see Braudel, 1995), it would be unfair to paint a picture of human society emerging in harmony with the environment until capitalism took shape. Nor are the 2.5 billion people living in smallholder or agricultural households always able to produce food without causing ecological problems (World Bank, 2007: 3). Nevertheless, the undeniable reality today is that northern agriculture creates particular sets of environmental challenges that not only remain unresolved, but are also set to become more serious if its geographical frontiers expand. It is to this issue that I now want to turn.

2.3 Amin's nightmare

In addition to the striking features discussed above, another prominent issue about northern agriculture refers to its impact on food producers around the world who do not use the land in this way. Note that the 2.5 billion living in smallholder or agricultural households (World Bank, 2007: 3) – many of them working tiny plots adjacent to new export-oriented monocultural zones exporting non-traditional crops – 'manage over 80 per cent of the world's estimated 500 million small farms and provide over 80 per cent of the food consumed in a large part of the developing world' (IFAD, 2013: 6). In Latin America, these peasant and small-scale farmers use 63 per cent of all farmland (Altieri, 2010: 120), and in sub-Saharan Africa they produce 70 per cent of the food supply (IFAD, 2013: 10).

What I want to suggest is that northern agriculture affects these food producers in some profound ways. And of particular importance is the growing gap between what the input-dependent farmers in northern agriculture can yield and what more traditional agricultural practices churn out. There are stark consequences emerging from this gap, which I refer to as 'Amin's nightmare' after Samir Amin, a prolific scholar whose work has explored the implications of capitalism for diverse people around the world, with a strong emphasis on the 'third world'. In an essay in 2003, Amin pointed out the following:

> Capitalist agriculture governed by the principle of return on capital, which is localized almost exclusively in North America, Europe, Australia, and in the Southern Cone of Latin America employs only a few tens of millions of farmers who are no longer peasants. Because of the degree of mechanization and the extensive size of the farms managed by one farmer, their productivity generally ranges between 1 to 2 million kilograms (2 and 4.5 million pounds) of cereals per farmer.
>
> In sharp contrast, three billion farmers are engaged in peasant farming. Their farms can be grouped into two distinct sectors, with greatly different scales of production, economic and social characteristics, and levels of efficiency. One sector, able to benefit from the green revolution, obtained fertilizers, pesticides, and improved seeds and has some degree of mechanization. The productivity of these peasants ranges between 10 000 and 50 000 kilograms (20 000 and 110 000 pounds) of cereals per year. However, the annual productivity of peasants excluded from new technologies is estimated to be around 1000 kilograms (2000 pounds) of cereals per farmer.
>
> The ratio of the productivity of the most advanced capitalist segment of the world's agriculture to the poorest, which was around 10 to 1 before 1940, is now approaching 2000 to 1! That means that productivity has progressed much more unequally in the area of agriculture and food production than in any other area.
>
> One can imagine that the food brought to market by today's three billion peasants, after they ensure their own subsistences, would instead be produced by twenty million new modern farmers. The conditions for the success of such an alternative would include: (1) the transfer of important pieces of good land to the new capitalist farmers (and these lands would have to be taken out of the hands of present peasant populations); (2) capital (to buy supplies and equipment); and (3) access to the consumer markets. Such farmers would indeed compete successfully with the billions of present peasants. But what would happen to those billions of people?

Might it be possible that three billion peasants could be replaced by 20 million capitalist farmers? In an influential report published in 2007, economists working for the World Bank called on governments around the world to focus on ways of promoting agricultural development (World

Bank, 2007). And in a crucial part of the report they pinpointed what they referred to as a 'yield gap' between what peasant producers manage to get out of the land and what might be achieved were more 'modern' agricultural practices introduced. For example, whereas the cereals yield per hectare in developed countries was around 5.5 tons in 2005 and had consistently grown from the 1960s to reach that level, in sub-Saharan Africa the yield was only around 1 ton per hectare and had hardly increased at all (World Bank, 2007: 67). In turn, this growing yield gap points to areas where land remains, from the perspective of these economists, under-used. Consequently, the report called on governments in those places to focus their efforts on closing the yield gap, such as by enabling more food producers to use inputs, machinery, and other innovations associated with northern agriculture.

I am struck by this idea of the yield gap and how it relates to Amin's nightmare. It seems to give a justification for pursuing the dramatic changes Amin warned about. For one thing, if the 20 million farmers Amin mentioned were to emerge, they would constitute a massive new market for the supplier firms. They will likely become tied into the networks of supermarkets and other retailers, which would further boost *their* chances of expanding. And from the perspective of the broker and supplier firms upstream of northern agriculture, as well as many of the corporations downstream of the farm, it seems as if it would make sense to pursue an expansion of northern agriculture's frontiers – thereby locking more and more people and places into the logic of the corporate food regime. Then there are private investors to think about – maybe even investors with no real track record, nor long-term interest, in the agricultural sector but who might consider it worthwhile to invest in land that can feed into the circuits of northern agriculture and potentially generate a return on investment greater than what is available in a capitalist economy defined today by low interest rates. Moreover, none of this is to even consider that a growing animal population means growing demand for animal feed, with the inevitable outcome that more and more cereals that could be used for human consumption are going to be processed into cattle feed. Cereal or soybean farmers in places such as the US or Europe will no doubt aim to meet much of this growing global demand for feed. But governments and businesses throughout the rest of the world will also want to see agricultural development step in. For them, therefore, growing meat consumption means growing markets, possibly more exports, and opportunities for expanding the area farmed by contemporary quasi-industrial methods.

Thus, the way our food system has emerged adds to the pressure to displace subsistence agriculture with capitalist northern agricultural practices; that is, the system's logic points to the realization of Amin's nightmare. Small-scale farmers, peasant producers, and others who remain outside of the circuits of northern agriculture loom large in present-day efforts to transform the global food economy. Indeed, as I now want to discuss, many of the conditions needed for northern agriculture to expand – for the corporate food regime to extend its grip over the food system

– have been laid down in the last few decades by some far-reaching shifts that overlap with the changes charted above and in Chapter 1.

2.3.1 Colonialism, development, and the foodscapes of neoliberalism

When Europeans (and then, much later, Japanese) colonialists began re-ordering the world from the fifteenth century onwards, they explored ways of re-making local agrarian systems in ways that would suit their interests, which in many instances entailed looking to satisfy demand from distant buyers. Colonialism therefore entailed finding ways of identifying and then taking the best land from the colonized. Plantations needed one type of land, and settler farms needed another. Colonial officials had to have an eye on water, on potential transportation networks, and on how they might secure a supply of labour. Colonial states and their officials had to imagine, map, and re-organize space – often using elaborate new laws such as the 1913 Natives Land Act in South Africa, which ultimately 'reserved' just a small portion of the country for the majority black population, while securing the rest for white farmers (e.g. see Beinart, 2001: 9–15; also Bernstein, 1996) – in ways that suited their interests and always at the expense of vast numbers of indigenous people. The specific crops grown varied spatially (e.g. bananas in one region, sugar in another) (on bananas, see Chapman, 2007; on sugar, see Abbott, 2009), and temporally (e.g. in Ceylon, present-day Sri Lanka, British planters experimented with growing coffee but eventually settled on growing tea; see Duncan, 2002); but one general pattern was to establish entirely new agricultural zones, relying on local as well as imported labour and transportation networks to bring the goods to local ports and then to distant markets. Experimentation was crucial in all this and it often had far-reaching consequences. In southern Africa, for example, maize seeds, which were imported from the Americas by the Portuguese, proved to be well suited to the soils, climates, and socio-economic ways of life, hence maize eventually became central to the agrarian economy throughout the region (McCann, 2005). Likewise, colonial rulers had to experiment with labour systems, ways of encouraging settlers to arrive, and how to manage crisis periods. In India, for instance, some of these experiments were horrific, as British rulers tested economic theories about state intervention amidst famines in 1876–1879 and 1896–1902 that killed up to 29.3 million people (Davis, 2002: 7).

Colonialism deposited other structures and relations, leaving complex ground on which newly independent states would have to realize their ambitions. Railways, for example, were laid according to what colonial rulers needed, not necessarily what was best for the indigenous population (e.g. see Griffiths, 1995: 181–186). Likewise, ports and urban areas were located and organized in ways that suited colonizers. Further, colonial rulers often established new hierarchies and new power relations between ethnic groups, tribes, or clans; relations that had to be negotiated after independence (e.g. see Mamdani, 1996).

The wave of independence movements which swept through colonized countries from the early 1950s had to build on those earlier investments and interventions. But independence also saw the deposition of new relations, practices, ideas, and technologies, many of which have shaped food production. Critically important here is what we might refer to as the 'development project' (McMichael, 2008a), which envisioned the creation of new post-colonial societies that would eventually achieve the living standards and material and social prospects akin to the great North American and European powers.

One part of this project focused on transforming agrarian sectors, especially via the so-called 'Green Revolution', which was introduced and expanded amidst rapid population growth and significant fears that food production would not keep pace with such growth, thereby leading to hunger or famine. The scientists and technicians behind the Green Revolution explored ways of breeding hybrid varieties for the main cereal crops (wheat, maize, and rice) and, without doubt, these new 'miracle' seeds boosted outputs. But merely planting miracle seeds in countries such as India, Pakistan, or the Philippines was never enough on its own; rather, the Green Revolution was based on the assumption that additional inputs would be used, especially inorganic fertilizer, pesticides, machinery, and water (Sage, 2012: 40) – this meant new markets were established for the producers of some of those inputs (Weis, 2007: 94). For many farmers and peasants, purchasing and then paying for inputs proved difficult, hence many fell into debt or simply saw their yields fall. Reflecting the belief that strong states should exert control over agrarian sectors via supports for agricultural modernization, protection from imports, and interventions such as agrarian reform (McMichael, 2008a), some governments stepped in to provide subsidies and supports for purchasing the inputs. However, government budgets did not always allow for sufficient purchases, and distributing subsidies and other inputs frequently created political tensions about who should benefit and why.

It is also worth noting here that, in many contexts, adopting Green Revolution ideas meant governments could avoid or give minimal attention to other causes of poverty and hunger, not the least of which was (and in many cases today, still is) about unequal distributions of land. In India, for example, the Green Revolution was used 'as a means of increasing food production without upsetting entrenched interests' (George, 1990: 184; in Weis, 2007: 94). An alternative strategy was to pursue land reform, that is, by strengthening tenure rights for peasants or redistributing land to meet the needs of the landless. This option was viewed by many, especially those on the left, as the key to reducing hunger and securing the rights of peasants. In contrast, the Green Revolution presented an alternative pathway to increasing food production without threatening the entrenched interests of landowners and elites.

Beyond the agrarian sector, the development project meant finding ways of achieving industrialization, expanding electricity production, creating hydroelectric projects, and building new road and port infrastructures (McMichael, 2008a). Consequently, many of the newly independent, but centralized and interventionist, states sought to borrow and invest in creating modern infrastructures. Banks in the US, Britain, and France were willing to lend enormous sums of money to post-colonial governments. They saw opportunities for returns on investment as well as the creation of entirely new markets that would, if things worked out as planned, eventually consist of new borrowers as well as new customers for other rich-world businesses. Such efforts to modernize and industrialize did not always work out. For every development 'star pupil' such as South Korea, whose society was transformed from a largely agrarian one to one of the world's most-industrialized countries (e.g. see Kay, 2002; Armstrong, 2007), there were numerous other cases where the 'development dream' failed entirely. Very few countries have actually managed to change their position in the international division of labour via industrializing and modernizing (on these 'persistent peripheries', see Dicken, 2011: 35–36). In part, the legacy of colonialism shaped what emerged. Many of the newly independent states got off to a bad start because they did not have enough people with the background, experience, and skills needed to run bureaucracies, institutions, and agencies (e.g. see Meredith, 2005: 277–278); colonial governments rarely wanted to develop much of an industrial infrastructure in their 'possessions', preferring instead that they remained mere producers of raw materials. Others found they had not only been dominated by colonial rulers in the guise of Governors-General reporting back to London, say, but also by the trading companies, the owners of plantations, and all of the other diverse actors who had a stake in ensuring that the newly independent society would continue to generate profits (Meredith, 2005: 153).

Rising government debt need not turn out to be a major problem if there is economic growth. But economic crises make debt repayments difficult and in the late 1970s and into the early 1980s a mixture of shifts and forces combined to make a mess of the development project. One was the formation of the Organization of the Petroleum Exporting Countries (OPEC) and the subsequent increase in oil prices, which forced oil importers to pay more for the fuel they had become accustomed to using. With rising fuel prices, African importers turned to international banks and borrowed so heavily that external debts increased from $6 billion to $38 billion between 1970 and 1980 (Meredith, 2005: 282). A second was falling prices for tropical agricultural commodities (Meredith, 2005: 375), caused in part by an over-supply – that is, too many places were producing too many of the same goods – but also by falling demand from the buyers of those commodities in rich-world countries which, in the late 1970s in particular, were going through a wave of economic crises. Finally, as part of a move to address economic difficulties at home, the US Federal Reserve rapidly increased interest rates in the early 1980s and this meant that the interest on borrowings to rich-world banks rose and made it almost impossible for indebted governments throughout the world to repay their loans (Brenner, 2006; Klein, 2007: 159). These three shifts amplified the task of overcoming the ruptures, tensions, problematic structures, and practices that colonialism introduced around the world.

With rising import bills, falling prices for exports, and rising interest rates on their debts, the development project essentially came to an end. In its place emerged a new 'neoliberal' vision for how all countries, not just in the post-colonial world, could develop. According to this neoliberal viewpoint, economic growth is more likely to occur if government finances are brought under control and government interventions minimized. The aim is to liberate businesses; to let them decide on how best to allocate resources. In the context of their difficulties making debt repayments, then, governments in newly independent states turned to the International Monetary Fund (IMF), which provides financial support in times of crisis, for example by making debt repayments but always also with an eye on pushing for reforms that are intended to ensure governments avoid getting into difficulty again. Particularly from the early 1980s onwards, it paired up with the World Bank to call for indebted governments to introduce reforms under the name of 'structural adjustment programs' (SAPs) (e.g. see Klein, 2007). These were radical disciplining and transformative *technologies* that entailed re-making and re-ordering political and economic structures, adjusting them, and inserting entirely new practices and relations. Ultimately, SAPs demanded currency devaluations, cutbacks in government spending such as on subsidies, and the creation of new revenue streams, not least via the privatization of state-owned assets. They entailed bold and often brutal attacks on living standards. They made a reality of the neoliberal viewpoint circulating within the IMF and World Bank – not without debate and opposition (on this, see Stiglitz, 2002) – that the state should stop intervening so much in the economy; that the government should be forced to try harder to balance its books in the immediate term rather than investing for the long term by accumulating debts; and that the economy as a whole should be oriented towards capitalizing on any advantage in export-oriented sectors.

Rather than investing in industrialization, therefore, as was central to the development project, the neoliberal approach claimed a country's best chance of paying back debts was to restructure the economy, especially by giving a boost to exporters, whether domestic-owned or foreign. In addition, instead of protecting domestic producers – agricultural or industrial – structural adjustment called for these 'market distortions' to be removed; hence quotas, tariffs, and subsidies were reduced or entirely dismantled, in part via SAPs but then also via the World Trade Organization, which left agricultural subsidies intact in the US and Europe (as I discussed in Chapter 1). In other words, structural adjustment introduced so-called 'free' trade policies, often leading to new bi- or multi-lateral trade agreements (on Latin

America's experience in this regard, see Green, 2003; Klein, 2007; on the situation in African countries, see Stiglitz, 2002), and the creation of numerous export processing zones where investors could access cheap, flexible workers.

With respect to food producers, these neoliberal ideas had far-reaching effects. As part of post-independence efforts to achieve higher food production via a strong interventionist state, many post-colonial governments had created agencies, associations, cooperatives, and other institutions that sought to shape or control agricultural production, and not least via controlling prices peasant food producers would be paid for their produce or would have to pay for inputs such as fertilizer (see Bernstein, 2005: 76–79). But so far as neoliberal orthodoxy was concerned, state-run structures distorted the market and created opportunities for corruption or political manipulation. They had to be taken apart; hence wherever the IMF and World Bank found them, these institutions were targeted and often dismantled. Although the idea was that food production would grow once these distortions were removed, such a reality did not always emerge. State-run structures often provided seed, fertilizer, chemicals, and other forms of support. Their demise left many producers in the lurch (e.g. on Guinea see Clapp, 1997; on Malawi, see Bezner Kerr, 2010).

Another outcome saw the emergence of export-oriented zones producing not food crops for domestic consumption, but rather new so-called 'non-traditional' crops for export, such as flowers, asparagus, baby carrots, strawberries, or mange tout (e.g. see Akram-Lodhi, 2013: 137–139). Often grown in enormous glass greenhouses on land once used to grow food for local consumption, these non-traditional agricultural exports (NTAEs) have boomed in the aftermath of structural adjustment and continue to be significant today. In Kenya, for example, the value of such exports increased throughout the 1980s and 1990s and by 2000 'Kenya's fresh vegetable trade exceeded $100 million, while the total combined exports of horticultural and floricultural products exceeded $250 million' (Jaffee *et al.*, 2011: 108), thereby contributing nearly two-thirds of Kenya's agricultural exports. Obviously, the owners of these exporting firms employ some local workers and they make profits, some of which will be re-invested locally. But whether these uses of land and water are the most appropriate for the various places in question is up for debate. In Kenya, 26 per cent of the population (just over 11 million people) were undernourished in 2013, yet still land is used to grow fresh cut flowers for consumers in Europe.

A final change flowing from structural adjustment was a new wave of food imports. The logic here was that consumers in Mexico, say, would be better off if they could purchase (subsidized, surplus) food from low-cost producers such as the US (on Mexico, see Hawkes, 2006). Prior to structural adjustment, most states limited food imports by erecting tariffs that protected domestic producers. The logic of structural adjustment was that tariffs and other quotas distorted markets and so they had to be removed (and as we saw in Chapter 1, the WTO Agreement on Agriculture has further curtailed the use of these instruments). Other supports to agricultural producers, such as low-cost credit, were also targeted (Bello and Baviera, 2010). These sorts of structural adjustments made it harder for domestic producers to compete against imports. The outcome for many countries has been a 'perilous' growing dependence on food imports (Weis, 2013: 75–81), which suits the producers and movers of food from the heartlands of northern agriculture, but places farmers and consumers in poor countries in a vulnerable position, especially in the context of food price volatility. In Guinea in West Africa, for example, imports of Asian rice surged in the aftermath of structural adjustments in the late 1980s. Rice imports were liberalized, thereby allowing traders to buy on the world market where rice could be purchased for less than it was sold for by local rice producers. Local production suffered as a result (Clapp, 1997). Similar outcomes emerged wherever local producers were confronted with imported crops produced using fertilizer and machinery, all while enjoying subsidies.

2.3.2 The land grab looms large

The development project was bound up with generating national food security via a strong interventionist state that subsidized and sought to protect domestic production. In contrast, the emergence of neoliberal ideas has entailed privatizing food security (Fraser, 2011), making it dependent not just on local food production and domestic prices but on the ability of buyers to afford global food prices. Furthermore, these developments have created what I think we should imagine as a weaker global south: one in which the scope for governments to *really* pursue changes that will benefit their populations has been seriously eroded. Just as the WTO has taken sovereignty away from national governments regarding trade rules (as I noted in Chapter 1), structural adjustment and the resulting commitment to neoliberal ideas – to openness, to limited intervention – chips away at the control governments have (even if those in office do not want to use them) to tackle deep and enduring social problems. And so this is what I meant when I said earlier that the conditions for Amin's nightmare to occur have been laid down over the last few decades. The overall food system is now framed by the corporate food regime; it is dominated by northern agricultural practices; and its internal relations – the relations between individual actors and their 'freedoms' to accumulate and then the market opportunities likely to open up to them, whether to produce feed, biofuels, or some other crop – suggest that dramatic changes are likely to occur in the open, porous global south.

Given the logics at work here, it just seems correct to conclude that land held and used by smallholders throughout the world is going to be increasingly targeted. In many places, what these shifts will mean is that some landowners (whether they are farmers or not) in tune with market opportunities will try to buy land from their neighbours or on the open market and gradually increase the scale of their operations, much as has happened in rich-world countries (Akram-Lodhi, 2012: 147). These landowners or farmers will create larger-scale businesses or farm units and will therefore be more likely to look for northern agricultural solutions to increase output and improve productivity. They might turn to GM seeds, mechanization, fertilizer, and so on. The spaces of northern agriculture will grow in size via endogenous processes.

A much more controversial issue, however, is when exogenous forces lead the way. In this regard, a striking recent development is the so-called 'global land grab' (e.g. see Zoomers, 2010). This idea of a land grab, which needs to be seen in the context of land dispossession during colonial times, tends to refer to agreements to exchange or lease land between governments in so-called 'developing' countries and foreign investors – which can include individuals and companies, but also sovereign wealth funds run by governments such as Saudi Arabia, the United Arab Emirates, and China. Many deals are large, covering up to 1.5 million hectares, such as land purchased in Sudan in 2009 by the Sayegh Group, a multinational conglomerate of companies (GRAIN, 2012). Other deals are smaller, sometimes involving only a few thousand hectares, such as the 4000 hectares that Liu Jianjun, a former Chinese government official, purchased in Uganda in 2008 (*Spiegel*, 2008). Many such land grabs do not actually lead to any changes – the World Bank has suggested that farming has only started on 21 per cent of the announced deals (Deininger and Byerlee, 2011: xiv) – but over the long term new agricultural zones will emerge and regardless of what size these deals are, it is clear that a significant amount of agricultural land has changed hands in recent years.

The outcome of this process is hard to fully predict. But there are certainly widespread fears that land grabs will place local food security at risk, especially when land that might once have been used to grow food is converted to produce biofuels or other export crops (e.g. see McMichael, 2012). Then there are concerns that the deals will exert pressure on water resources, not least if landowners take a short-term view and look to squeeze as much use out of the land as possible; in fact, it seems to some as if land grabs are increasingly about speculative 'water grabs', that is, where investors are more interested in access to water resources than the land itself (Franco and Kay, 2012).

According to a report published by the World Bank (Deininger and Byerlee, 2011: 89–90), the greatest potential for displacement to occur is in

> sparsely populated countries—such as the Democratic Republic of Congo, Mozambique, Sudan, Tanzania, and Zambia—with large tracts of land suitable for rain-fed cultivation (in areas of sufficient precipitation) but also a large portion of smallholders who only achieve a fraction of potential productivity.

If governments can create the right 'institutional framework [and] complementary infrastructure' (Deininger and Byerlee, 2011: xxxvii), they will be well placed to lure in outside investors. Crucially, although the World Bank's experts state that such deals need to include 'recognition and respect for existing land rights' (Deininger and Byerlee, 2011: 91), as well as full public disclosure of contractual arrangements and oversight of compliance, it is not too difficult to imagine how the opposite can easily occur. As Vermeulen and Cotula (2010) put it, land deals tend to develop between government agencies and investors and therefore 'over the heads of local people'. Opaque land deals between governments with poor human rights records and the sovereign wealth funds of other governments with equally questionable records, or with private investors looking to earn vast profits: it is easy to see how it can go wrong for poor, marginalized, perhaps even illiterate people who stand to lose land they have been using for subsistence purposes, or even for the market.

In short, the land grab points towards one way that Amin's nightmare might take shape. The 2.5 billion peasants and smallholders who populate the spaces of Amin's nightmare across the so-called global south have experienced all of the various waves of investments and interventions occurring in the last few decades: colonialism, the Green Revolution and developmentalism, structural adjustment, the growth of NTAEs, and rising food import dependence. Nevertheless, many continue to use the land much as they might have done a few decades ago, albeit with the exception that today their potential markets for sales are now subject to competition from food imports. They often occupy land right next door to the export-oriented sector, perhaps even using the same water resources, taking in the same view. Their co-existence today equates to a 'bifurcated agrarian sector' (Akram-Lodhi *et al.*, 2008), with the export sector producing for sale, while the peasant produces for use. One looks to consumers in the richer countries in Europe, North America, or Asia; the other mainly focuses on local opportunities in addition to waged work that might boost their incomes, in part because of rising food imports.

Amin was right to ask what will happen to those people. I cannot provide an answer, but there seems little doubt that their prospects seem dire. Some will look for paid work, perhaps on a part-time basis on plantations or working for wealthier export-oriented farmers, although there is downward pressure on wages as employers seek to cut costs and meet the requirements of buyers (World Bank, 2007: 213). Many will no doubt just leave the land. They might look to find waged work in the city, although they might only find a grim 'wageless life' (Denning, 2010) in what Mike Davis (2006) has referred to as our contemporary 'planet of slums'. And others still will do what many in Central America do: migrate to richer countries, such as the US, and hope to find employment, however casual, and hopefully send money home. As the rest of the materials in this chapter demonstrate, the pressures exerted by northern agriculture's development are intense.

However, as we will see in the final section, any notions we might have here that peasants are passive or entirely powerless need to be kept in check.

2.4 Case study: soy in Brazil and Paraguay

The geographical expansion of northern agriculture is never going to be straightforward or smooth. Nor is it going to be a benign process, whether we consider its impact on people or the wider environment. Indeed, if its expansion has the energy to displace the world's three billion peasants, it will be a horrible process for many. The cases of Brazil and Paraguay are illuminating here. Soy is an incredibly powerful, pliable plant which produces a bean rich in protein. Processors have found ways of using the entire soybean, hence soy-based products can be found in chocolate bars, baked goods, milk, and even protein drinks. But by far the most significant use of soybeans is for the production of feed for livestock such as cattle, pigs, or chickens, and for farmed fish, an area of the global food economy that has been rapidly growing in recent years. As diets globally have incorporated more meat and fish, the demand for soybeans has soared, resulting in transformations in the places best-suited to growing the crop.

Although the US has been the leading producer since the 1960s, a significant production zone today is the Amazonian 'Republic of Soy' (Holt-Giménez and Shattuck, 2010: 79), an area of around 50 million hectares (twice the size of the UK) of soybean farms shared between Argentina, Brazil, and Paraguay. Soybean production in Brazil took off from the 1970s, driven by government support for territorial expansion and public and private sector investments in processing facilities (see Patel, 2007: 165–214). It is now a soy giant, producing as much as, and in some years more than, the US. Much of Brazil's production has involved cutting into the rainforest, with all the attendant environmental concerns, but recent government policies have slowed down the rate of deforestation (Revkin, 2014), even as soy production has increased. To a large extent, its booming production figures stem from tighter processes, reflecting precisely the sort of focus on productivity that we would expect from northern agriculture.

Soybean production emerged later in Paraguay. It is now the world's sixth biggest producer (after the US, Brazil, Argentina, China, and India), reflecting a sowing surface that has expanded from 1.3 million hectares in 2000–2001 to 2 million hectares in 2007–2008 (García-López and Arizpe, 2010). This growth has been led by Paraguayan farmers with access to credit who have purchased land from less competitive farmers, often those operating on a small scale. But farmers from Brazil and Argentina have also moved into Paraguay, generating rising land values and competition for new land. The result is that around 100 000 small-scale farmers have lost access to land, often via evictions by landowners (Howard, 2009). As such, the expanding area under soy exacerbates the problem of peasant access to arable land:

> Paraguay is the country with the greatest concentration of land ownership in Latin America: 77 per cent of arable land is owned by one per cent of the population. Small farmers, who represent 40 per cent of the population, own just five per cent of all farmland.
>
> (García-López and Arizpe, 2010: 199)

In addition, the mechanized production now pursued by soy farmers results in fewer jobs becoming available in the agricultural sector, impoverishing those needing work or forcing them to consider moving to urban areas.

The soybean boom has also altered environments:

> A biologically diverse Interior Atlantic Forest once covered 85 percent of Eastern Paraguay. Intermingled with the necessary shade and fruit-bearing trees of the forest, farmers grew diverse crops and raised a variety of livestock. However, today only 5–8 percent of that forest remains.
>
> (Howard, 2009: 39)

It is also crucial to note that soybean farms use toxic chemicals that poison groundwater, the air, and deposit toxins in the soil. Many people have been forced to relocate because soy expansion is wrapped up with the use of agricultural chemicals, often sprayed onto the land by plane. Indeed, those

> who live next to the soy fields have been driven away by the chemicals, which kill their crops and animals and cause illnesses. ... Children have died as a result of coming into direct contact with chemical clouds. In most cases, farming families are isolated and lack medical or monetary resources, so their complaints go undocumented.
>
> (Howard, 2009: 40, 41)

Perhaps not surprisingly, all of these seemingly relentless, runaway changes are supported – and called for – by some of the global food economy's largest, most powerful transnational corporations. Archer Daniels Midland, Bunge, and Cargill are all present in Paraguay, selling seeds (including GM seeds) and buying from farmers. Cargill, for example, owns 'the largest processing capacity for oil seeds in Paraguay' (Holt-Giménez and Shattuck, 2010: 82; also see Newell, 2008). Clearly, Paraguay occupies a position within the corporate food regime that satisfies the expanding need for markets among supplier firms and the demand for further output among the buyers and processors, not least in the context of rising meat consumption, which soy production in places such as Paraguay is largely intended to meet. In turn, it is necessary to view changes in a place such as Paraguay not just as reflections of processes originating in the local area, but also as stemming from the internal tensions of northern agriculture and the broader corporate food regime. In Paraguay, we can see how northern agriculture expands geographically and the types of changes it brings about.

2.5 In what sense oppressive?

Without doubt, there are numerous forms of oppression in this second stage in the life of food. There is the exploitation of farm workers, the marginalization of peasant forms of agriculture, the powerlessness of many commercial farmers who feel as if they are stuck on the treadmill of synthetic inputs, and then the cultural imperialism of a certain type of agricultural practice coming to dominate all others throughout the world. More broadly, we might consider how women working the land suffer multiple forms of oppression at the same time (e.g. on rural women's experience in Bangladesh, see Ahmed *et al.*, 2004). My focus here, though, is on violence.

Recall that Iris Young (1990) identified violence as a form of oppression because some social groups are intimidated, harassed, humiliated, or threatened with (and often subjected to) beatings or rape. Throughout the world, the actual production of food on the land is intimately wrapped up with violence, both historically and in the contemporary period. In a feudal context, landowners used violence as a means to compel serfs to work the land (Meiksins-Wood, 2002), and in some contexts around the world such landlord–serf relations persist (Akram-Lodhi, 2013). The origins of capitalism entailed sweeping away feudal relations, often in a violent manner as lands were enclosed and peasants forced to move to urban areas. Colonialism, too, entailed violent enclosure, land clearances, repression of indigenous peoples, and then numerous forms of violent governance of plantation workers and farm labourers (e.g. on Ceylon, present-day Sri Lanka, see Duncan, 2002). We can trace the violent origins of the global food economy wherever food is produced (see also Carney, 2001).

In the contemporary period, it might be nice to imagine that the violent ways of the past have been swept away. Certainly, in the heartlands of northern agriculture, laws and policing might manage to place a limit on the violence landowners or farmers can enact, although this is not to say violence has ended. But we should not ignore the violence – the vast endless flows of blood – enacted against the billions of animals living in CAFOs and brought to their deaths in slaughterhouses. And nor can we turn away from the violence unfolding in places that play a major role in the production of the food we eat. Consider Paraguay again. Landowners – those with capital to invest, with agribusiness interests looking to increase their access to profits – have been more than willing to use violence. Some soy farmers hire private security forces to bully adjacent land users to sell up; peasant-led protests have been quashed with help from the government's security services; and threats and intimidation have been followed up with killings (Howard, 2009). In so far as northern agricultural practices have expanded in Paraguay, it is necessary to view that expansion as moved forward via violent means. In neighbouring Brazil, too, violence against landless workers, some of whom have moved to occupy land (a theme explored in more detail in the next and final section of the chapter), remains common. The expansion of cattle ranching and commercial farms modelled on northern agricultural practices has generated numerous tensions between newcomers and indigenous peoples, many of whom have suffered decades of state-led interventions designed to open Brazil to settlers. Elsewhere, the Amazon Land War in the South of Pará – an area with some of the most pronounced land inequality in Brazil (Simmons *et al.*, 2007: 583) – has pitted wealthy landowners against peasants and landless workers. Some landowners are now so willing to assassinate their opponents that prices for targeted murders – of squatters or lawyers, priests, or politicians who support them – are well known (Simmons *et al.*, 2007: 583). Many of the landless who have now organized themselves to undertake direct action land reform now encounter organized militias paid for by the landowners, such as the *Uniao Democratica Ruralista*, which is 'dedicated to the defense of private property and the large-holdings' (Simmons *et al.*, 2007: 584), and which frequently represses the landless (see also Hammond, 2009).

The upshot of all this is that violence – not just money with which to purchase land – is central to Amin's nightmare becoming a reality. Those who use the land, after all, do not simply attach monetary values to it: whether white commercial farmers in post-Apartheid South Africa (Fraser, 2008, 2010) or peasant producers in northern Thailand (e.g. see McMichael, 2008b: 55 fn.13), land has a strong emotional pull, has meaning beyond money, and is something lots of people consider worth fighting for, not least if having land is all that stands between needing to move to an informal settlement nearby, or a distant city. For those with wealth – those with the money to pay militias or private security firms – using violence to make possible their ambitions might seem a reasonable action, especially if laws and policing are weak. Via dominating others, farmers with credit and finance and with the connections and networks to ensure their products can reach markets, have brought about rapid transformations in their respective areas but at the expense of others (and with far-reaching consequences for the environment, too). These scenarios are exactly what we should expect, given Amin's nightmare; and in places such as sub-Saharan Africa, where land grabs are under way, the trend of richer farmers pushing out poorer land users looks set to be repeated, possibly on an expanded scale.

2.6 Resistance: occupying to craft alternative agricultures

The discussion up to now has been rather grim. There is a general direction of change and it tends to point towards the expansion of northern agricultural practices, often via violent methods. But it is essential to note here – as it is throughout this book – that there are lots of ways in which people across the world are trying to stand up against the dominant direction of change. In some parts of the world, for example, there are efforts to re-imagine agriculture as something that need not require synthetic inputs, as something that can be organic (e.g. see Guthman, 2004). Elsewhere, there are cases of agricultural workers using strikes to push for safer working conditions, better wages, and recognition (e.g. see Mitchell, 2013). Then there are cases of some consumers boycotting agricultural produce from places where practices are known to be particularly oppressive (e.g. on the role of consumer boycotts in campaigns against apartheid South Africa, see Gurney, 2000). But instead of focusing on any of these examples, the mode of resistance I want to dwell upon here is the tactic of occupying land.

Occupying has become a prominent mode of resistance in recent years, especially in the aftermath of the Occupy Wall Street movement, which sought to create an autonomous space of resistance in the heart of New York's financial district (Harvey, 2012). But the tactic has a much longer lineage and broader geography. In places where land is unequally distributed, such as Brazil and other countries in Latin America, peasant movements have looked to occupy land as a way to grow food, but also to highlight their plight (e.g. Moyo and Yeros, 2005a). Elsewhere, in Zimbabwe in the late 1990s and then in some parts of South Africa in the early 2000s, land invasions were used to seize land from white landowners (Moyo and Yeros, 2005b; on land occupations in Asia as a whole, see Aguilar, 2005; on India, see Pimple and Sethi, 2005: 248–251; in Latin America, see Veltmeyer, 2005).

Probably no other movement has achieved as much success via occupying than Brazil's MST (after their name in Portuguese, *Movimento dos Trabalhadores Rurais Sem Terra*: Landless Rural Workers' Movement) (e.g. see Wolford, 2003, 2004). After identifying possible sites, the MST mobilizes large groups of landless workers to invade land, establish a settlement, begin growing, and then defend the new location against landowners and their 'security' forces. Brave moves. In doing so, the MST looks to improve the livelihoods of their members; but there is no doubt the movement also seeks to highlight the maldistribution of land in Brazil, with some estimates suggesting that the percentage of the total area occupied by the 10 per cent of largest properties is approximately 78 per cent (Oxfam, 2008).

In turn, legitimacy the MST has earned outside Brazil has enabled it to influence transnational agrarian movements, especially Vía Campesina, which is made up of about 150 local and national organizations representing 200 million farmers from 70 countries in Asia, Europe, Africa, and the Americas (see Desmarais, 2007; also Wittman *et al.*, 2010). Vía Campesina 'defends small-scale sustainable agriculture as a way to promote social justice and dignity [and] strongly opposes corporate driven agriculture and transnational companies that are destroying people and nature' (La Vía Campesina, 2011); and opposes the current direction of policy development regarding food. Its members are often present at trade negotiations, UN conferences, and at national conventions or demonstrations across the world. It has an activist stance that tries to counter the dominant arguments about how the global economy should connect with food and agriculture. But it is also important to note that Vía Campesina is at the forefront of

Figure 2.1 MST supporters in Brazil (source: estudio Maia/Shutterstock).

constructive calls for an alternative type of agriculture, one that remains globally oriented but with the aim of eliminating rather than, say, 'addressing' hunger. And the centrepiece of its argument is the notion of 'food sovereignty'.

In contrast to the notion of 'food security', which asks whether people are likely to suffer food shortages, the emphasis on *sovereignty* focuses attention on the need for small-scale farmers to have rights over the conditions under which they can produce food. It is, as Vía Campesina notes, about ensuring 'the right of peoples to healthy and culturally appropriate food produced through sustainable methods *and* their right to define their own food and agriculture systems' (La Vía Campesina, 2011). In a world in which food sovereignty is protected, a model would emerge

> of small scale *sustainable* production benefiting communities and their environment [and which] puts the aspirations, needs and livelihoods of those who produce, distribute and consume food at the *heart* of

> food systems and policies[,] rather than the demands of markets and corporations.
>
> (La Vía Campesina, 2011)

The emphasis on sustainable food production means 'the rights to use and manage lands, territories, water, seeds, livestock and biodiversity [should be] in the hands of those who produce food and not of the corporate sector' (La Vía Campesina, 2011), because the corporate sector puts profits before sustainability. Furthermore, and in contrast to the trade networks that move certain foodstuffs all across the world to cater to an artificial 'global summertime' market, food sovereignty 'prioritizes local food production and consumption [which] gives a country the right to protect its local producers from cheap imports and to control production' (La Vía Campesina, 2011). Again, this is in contrast to the argument that so-called 'free trade' should be enforced globally, an argument that strips small-scale farmers and peasants of their rights to protect forms of production that might run counter to the interests of TNCs.

The argument for food sovereignty demands a cut in agricultural subsidies in places such as the US and Europe and leads to the conclusion that the WTO's Agreement on Agriculture should be scrapped in favour of a new agreement that prioritizes food sovereignty over corporate power. It also advances the claim that genuine agrarian reform is central to the food sovereignty vision, that is, agrarian reform should protect peasant farmers from evictions and grant them land to produce the food they need (see Borras, 2008). Given that so much land globally is still owned by a small minority of landowners, land reform is a necessary stage in the creation of food sovereignty: in other words, redistributing land to those who can use it in a more sustainable way over the long-term to generate food. Ultimately, the argument is that small-scale farming can be more sustainable than modern, industrial agriculture based on a treadmill of technological fixes. For example, there is research suggesting that small-scale farming can be more productive per unit of land than monocultural food production. In Mexico, for example, Miguel Altieri (2010: 122) notes that:

> A large farm may produce more corn per hectare than a small farm in which the corn is grown as part of a polyculture that also includes beans, squash, potato, and fodder. Yet, for smallholder polycultures, productivity in terms of harvestable products per unit area is higher than under large-scale sole cropping with the same level of management. Yield advantages can range from 20 to 60 percent, because polycultures reduce losses due to weeds, insects and disease and make more efficient use of water, light and nutrients. In Mexico, a 1.73 hectare plot of land has to be planted with maize monoculture to produce as much food as 1 hectare planted with a mixture of maize, squash and beans. In addition, the maize–squash–bean polyculture produces up to 4 tons per

> hectare of dry matter for plowing into the soil, compared with 2 t
> a maize monoculture.

A major part of this inspirational vision for how food can be produced, without displacing peasants and without impacting too negatively on the environment, maps onto some other recent developments. In particular, the idea behind food sovereignty is bolstered by the concept of *agroecology*, which stems from overlaps between two scientific disciplines: agronomy and ecology. Agroecology seeks to apply what we know from ecology to agricultural systems, but in ways that focus on creating sustainable systems, rather than just productive or profitable systems. It therefore differs quite considerably from what has tended to dominate the science of agronomy, which has been heavily influenced by a productivist ethos focusing on getting more out of any given piece of land. Whereas the latter might look to farm land intensively, agroecology looks to farm land extensively but in a sustainable manner. There are other aspects, too. Here is what the United Nations Special Rapporteur on the right to food, Olivier De Schutter, says about it in a recent report to the UN (UN, 2010: 6):

> As a set of agricultural practices, agroecology seeks ways to enhance agricultural systems by mimicking natural processes, thus creating beneficial biological interactions and synergies among the components of the agroecosystem. It provides the most favourable soil conditions for plant growth, particularly by managing organic matter and by raising soil biotic activity.

The focus is on reducing waste by recycling nutrients and energy and cutting reliance on inputs. But agroecology also looks for ways to let species such as crops and livestock bounce off of one another in positive ways, hence biodiversity rather than monoculture is central to the vision. Agroecology is about creatively pursuing interactions that deliver productivity all across the agricultural system, rather than simply focusing on individual species (e.g. see Pretty, 2009). Some of the main elements include: maintenance or introduction of agricultural biodiversity; integrated nutrient management; water harvesting in dryland areas; integration of livestock into farming systems, such as dairy cattle, pigs, and poultry, to provide a source of protein to the family, as well as a means of fertilizing soils; resource-conserving, low-external-input techniques and enhancing on-farm fertility production to reduce farmers' reliance on external inputs and state subsidies (Pretty, 2009). All of these ideas revolve around improving yields in clever ways but without increasing reliance on technologies. In a way, this is 'smart farming', as the following quote from a UN report (UN, 2010: 8–9) makes clear:

> Sometimes, seemingly minor innovations can provide high returns. In Kenya, researchers and farmers developed the 'push–pull' strategy to

> control parasitic weeds and insects that damage the crops. The strategy consists in 'pushing' away pests from corn by inter-planting corn with insect-repellent crops like Desmodium, while 'pulling' them towards small plots of Napier grass, a plant that excretes a sticky gum which both attracts and traps pests. The system not only controls pests but has other benefits as well, because Desmodium can be used as fodder for livestock. The push–pull strategy doubles maize yields and milk production while, at the same time, improving the soil. The system has already spread to more than 10,000 households in East Africa by means of town meetings, national radio broadcasts and farmer field schools.

Supporting all these efforts is output from the International Assessment of Agricultural Knowledge, Science and Technology for Development (IAAKSTD), an extensive assessment of agriculture in the world run by hundreds of the world's foremost scientists. Its report, Agriculture at a Crossroads, suggests that a form of sustainable small-scale agriculture does a better job of supplying food in many of the poorer parts of the world than the type of northern agriculture I have charted in this book (Pretty, 2009). Rather than seeing Amin's nightmare unfold, activists in networks and movements such as Vía Campesina, as well as allied agroecologists and other scientists, argue for ways of creating a food system that provides food producers with a degree of sovereignty over the way they lead their lives.

The significance of what I have discussed here is this: in the same space-time that corporations up- and downstream of agriculture have become so profitable and powerful, hundreds of millions of people are effectively priced out of the food economy. Moreover, the same processes, policies, and practices that make it possible for corporations to profit in an expanding and globalizing marketplace exert pressure on millions of peasants and small-scale farmers to use the land in unsustainable ways, and even to abandon it. Given these developments, critics of the agricultural system have tried to look towards an alternative form of globalization that is oriented towards eliminating hunger and improving the lives of those who work the land. Instead of a corporate-friendly globalization, the call is for a type of globalization of food and agriculture that is, if you will, people-friendly, democratic, and anti-corporate. Via the act of occupying – as the MST has demonstrated – a debate has emerged in which new ideas have bounced off of one another and fed into a growing awareness of what a world *without* northern agriculture could be like. But occupy as a mode of resistance here is not just about taking over land. Rather, it is also about trying to occupy a place wherever debates about food production occur. Ideas such as food sovereignty and agroecology challenge dominant ideas about food today: they challenge the almost taken-for-granted wisdom that TNCs know best, that larger-scale farms are necessarily more efficient users of the land, or that countries are better off exporting non-traditional crops.

What else matters here?

I have focused on occupy as a mode of resistance in this stage in the life of food. If you were to examine another form of resistance, what might you reveal?

I have only touched on the idea that greater democratic control of our food system might be something we should look to achieve in the future. How might this occur? How can markets – capitalist firms – be brought under democratic control?

There is, without question, so much more to the global land grab story than I have covered. For one thing, its controversial character has to be seen in the light of a colonial past. Choosing a country where the land grab matters, just how do critics 'frame' things with respect to their colonial history?

You decide

- Imagine that government regulation changed the orientation of northern agriculture so that it became more sustainable, less focused on producing 'cheap' food, and less input-dependent. Would we have enough food in the world?
- If land grabs do result in closing the yield gap, will they be worth it?
- If the future for those caught up in Amin's nightmare is a life in urban slums and ultimately relying on imported food, what will this mean for global society? What new challenges will confront us in the next few decades?

Suggested reading

As my references clearly indicate, the books by Clapp (2012), Patel (2007), Sage (2012), and Weis (2007) are fantastic resources that anyone interested in the materials here must consult. I cannot speak highly enough of them. Also worth seeing are Akram-Lodhi (2013), Bernstein (2010), Carolan (2011), and Weis (2013).

On colonialism generally, *Late Victorian Holocausts* by Davis (2002) is an excellent resource. So are Arrighi (1994) and Galeano (1973).

For a critical and sharp analysis of neoliberalism, see Harvey (2005). And on the rationale, nature, and effects of structural adjustment programmes see Klein (2007), Green (2003), and Stiglitz (2002).

On land grabs, an influential paper is Zoomers (2010). She identifies seven processes driving land grabs, unlike the three I focus on. *The Journal of Peasant Studies* has numerous other useful articles on land grabs. McMichael (2012) is also worth seeing. Also important is the *Journal of Agrarian Change*, which has several papers and even special issues on land grabbing.

Note

1 One hectare is 100 m^2.

References

Abbott, E. (2009) *Sugar: A Bittersweet History*. London: Duckworth Overlook.

Aguilar, F. (2005) Rural Land Struggles in Asia: Overview of Selected Contexts. In: Moyo, S. and Yeros, P. (eds), *Reclaiming the Land: The Resurgence of Rural Movements in Africa, Asia and Latin America*. London: Zed Books, pp. 209–234.

Ahmed, M.K., van Ginneken, J., Razzaque, A., and Alam, N. (2004) Violent Deaths Among Women of Reproductive age in Rural Bangladesh. *Social Science & Medicine*, 59(2), 311–319.

Akram-Lodhi, A.H. (2012) Contextualising Land Grabbing: Contemporary Land Deals, the Global Subsistence Crisis and the World Food System. *Canadian Journal of Development Studies/Revue canadienne d'études du développement*, 33(2), 119–142.

Akram-Lodhi, A.H. (2013) *Hungry for Change: Farmers, Food Justice, and the Agrarian Question*. Halifax: Fernwood.

Akram-Lodhi, A.H., Kay, C., and Borras, S.M. Jr. (2008) The Political Economy of Land and the Agrarian Question in an Era of Neoliberal Globalization. In: Akram-Lodhi, A.H. and Kay, C. (eds), *Peasants and Globalization: Political Economy, Rural Transformation and the Agrarian Question*. London: Routledge, pp. 214–238.

Altieri, M.A. (2010) Scaling Up Agroecological Approaches for Food Sovereignty in Latin America. In: Wittman, H., Desmarais, A.A., and Wiebe, N. (eds), *Food Sovereignty: Reconnecting Food, Nature and Community*. Oxford: Pambazuka, pp. 120–133.

Amin, S. (2003) World Poverty, Pauperization & Capital Accumulation [online]. *Monthly Review*, 55(5). Available at: http://monthlyreview.org/2003/10/01/world-poverty-pauperization-capital-accumulation (accessed 22 January 2016).

Armstrong, C.K. (2007) *The Koreas*. London: Routledge.

Arrighi, G. (1994) *The Long Twentieth Century: Money, Power and the Origins of Our Times*. London: Verso.

Beinart, W. (2001) *Twentieth-Century South Africa*. Oxford: Oxford University Press.

Bello, W. and Baviera, M. (2010) Capitalist Agriculture, the Food Price Crisis & Peasant Resistance. In: Wittman, H., Desmarais, A.A., and Wiebe, N. (eds), *Food Sovereignty: Reconnecting Food, Nature & Community*. Oxford: Pambazuka, pp. 62–75.

Bernstein, H. (1996) South Africa's Agrarian Question: Extreme and Exceptional? *Journal of Peasant Studies*, 23, 1–52.

Bernstein, H. (2005) Rural Land and Land Conflicts in Sub-Saharan Africa. In: Moyo, S. and Yeros, P. (eds), *Reclaiming the Land: The Resurgence of Rural Movements in Africa, Asia and Latin America*. London: Zed Books, pp. 67–101.

Bernstein, H. (2010) *Class Dynamics of Agrarian Change*. Halifax: Fernwood.

Bezner Kerr, R. (2010) Unearthing the Cultural & Material Struggles over Seed in Malawi. In: Wittman, H., Desmarais, A.A., and Wiebe, N. (eds), *Food Sovereignty: Reconnecting Food, Nature & Community*. Oxford: Pambazuka, pp. 134–151.

Borras, S.M. Jr. (2008) La Vía Campesina and its Global Campaign for Agrarian Reform. *Journal of Agrarian Change*, 8, 258–289.

Braudel, F. (1995) *The Mediterranean and the Mediterranean World in the Age of Philip II*. Berkeley, CA: University of California Press.

Brenner, R. (2006) *The Economics of Global Turbulence: The Advanced Capitalist Economies from Long Boom to Long Downturn, 1945–2005*. London: Verso.

Carney, J. (2001) *Black Rice: The African Origins of Rice Cultivation in the Americas*. London: Harvard University Press.

Carolan, M. (2011) *The Real Cost of Cheap Food*. Abingdon: Earthscan.

Carson, R. (1962) *Silent Spring*. New York, NY: Houghton Mifflin.

Chapman, P. (2007) *Bananas: How the United Fruit Company Shaped the World*. Edinburgh: Canongate.

Clapp, J. (1997) *Adjustment and Agriculture in Africa: Farmers, the State, and the World Bank in Guinea*. Basingstoke: Macmillan.

Clapp, J. (2012) *Food*. Cambridge: Polity.

Davis, M. (2002) *Late Victorian Holocausts*. London: Verso.

Davis, M. (2006) *Planet of Slums*. London: Verso.

Deere Corp. (2014) *Annual Report* [online]. Available at: http://investor.deere.com/files/doc_financials/annual_reports/2015/2014_annual_report_v001_g1fx1i.pdf (accessed 22 January 2016).

Deininger, K. and Byerlee, D. (2011) *Rising Global Interest in Farmland: Can It Yield Sustainable and Equitable Benefits?* Washington, DC: World Bank.

Denning, M. (2010) Wageless Life. *New Left Review*, 66 (November–December), 79–97.

Desmarais, A.A. (2007) *La Vía Campesina: Globalization and the Power of Peasants*. London: Pluto.

Dicken, P. (2011) *Global Shift* (6th edn). London: Sage.

Duncan, J. (2002) Embodying Colonialism? Domination and Resistance in Nineteenth-Century Ceylonese Coffee Plantations. *Journal of Historical Geography*, 28(3), 317–338.

Earth Policy Institute. (2014) *Meat Consumption per Person in the United States, 1960–2013* [online]. Available at: www.earth-policy.org/?/data_center/C24/ (accessed 22 January 2016).

Environmental Protection Agency. (2014) *Major Crops Grown in the United States* [online]. Available at: www.epa.gov/oecaagct/ag101/cropmajor.html (accessed 22 January 2016).

Food and Agriculture Organization. (2014) *The State of Food and Agriculture, 2014: Innovation in Family Farming* [online]. Rome: Food and Agriculture Organization of the United Nations. Available at: www.fao.org/3/a-i4040e.pdf (accessed 13 April 2016).

Franco, J. and Kay, S. (2012) *The Global Water Grab: A Primer* [online]. Available at: www.tni.org/primer/global-water-grab-primer?context=69566 (accessed 22 January 2016).

Fraser, A. (2008) White Farmers' Dealings with Land Reform in South Africa: Evidence from Northern Limpopo Province. *Tijdschrift voor Economische en Sociale Geografie*, 99(1), 24–36.

Fraser, A. (2010) The craft of scalar practices. *Environment and Planning A*, 42(2), 332–346.

Fraser, A. (2011) Nothing but its eradication? Ireland's Hunger Task Force and the Production of Hunger. *Human Geography*, 4(3), 48–60.

Galeano, E. (1973) *Open Veins of Latin America: Five Centuries of the Pillage of a Continent*. London: Monthly Review Press.

García-López, G.A. and Arizpe, N. (2010) Participatory Processes in the Soy Conflicts in Paraguay and Argentina. *Ecological Economics*, 70(2), 196–206.

George, S. (1990) *Ill Fares the Land: Essays on Food, Hunger, and Power*. London: Penguin.

GRAIN. (2012) *Global Land Deals Dataset* [online]. Available at: www.grain.org/attachments/2453/download (accessed 22 January 2016).

Green, D. (2003) *Silent Revolution: The Rise and Crisis of Market Economics in Latin America*. New York, NY: Monthly Review Press.

Griffiths, I.L.L. (1995) *The African Inheritance*. London: Routledge.

Gurney, C. (2000) 'A Great Cause': The Origins of the Anti-Apartheid Movement, June 1959–March 1960. *Journal of Southern African Studies*, 26(1), 123–144.

Guthman, J. (2004) *Agrarian Dreams: The Paradox of Organic Farming in California*. Berkeley, CA: University of California Press.

Hammond, J.L. (2009) Land Occupations, Violence, and the Politics of Agrarian Reform in Brazil. *Latin American Perspectives*, 36(4), 156–177.

Harvey, D. (2005) *A Brief History of Neoliberalism*. Oxford: Oxford University Press.

Harvey, D. (2012) *Rebel Cities: From the Right to the City to the Urban Revolution*. New York, NY: Verso.

Hawkes, C. (2006) Uneven Dietary Development: Linking the Policies and Processes of Globalisation with the Nutrition Transition, Obesity, and Diet-Related Chronic Diseases. *Globalization and Health*, 2(4), doi: 10.1186/1744-8603-2-4.

Henry, T.A. (2005) Sanitizing empire: Japanese articulations of Korean otherness and the construction of early colonial Seoul, 1905–1919. *Journal of Asian Studies*, 64, 639–675.

Holt-Giménez, E. and Shattuck, A. (2010) Agrofuels & Food Sovereignty: Another Agrarian Transition. In: Wittman, H., Desmarais, A.A., and Wiebe, N. (eds), *Food Sovereignty: Reconnecting Food, Nature & Community*. Oxford: Pambazuka, pp. 76–90.

Howard, A. (2009) The Campesino Struggle for Sustainable Agriculture in Paraguay [online]. *Monthly Review*, 61(2). Available at: http://monthlyreview.org/2009/06/01/saying-no-to-soy-the-campesino-struggle-for-sustainable-agriculture-in-paraguay (accessed 22 January 2016).

IFAD (International Fund for Agricultural Development). (2013) *Smallholders, Food Security, and the Environment* [online]. Available at: www.unep.org/pdf/SmallholderReport_WEB.pdf (accessed 22 January 2016).

Jaffee, S., Henson, S., and Diaz Rios, L. (2011) *Making the Grade: Smallholder Farmers, Emerging Standards, and Development Assistance Programs in Africa*. Washington, DC: World Bank.

Kay, C. (2002) Why East Asia Overtook Latin America: Agrarian Reform, Industrialisation and Development. *Third World Quarterly*, 23(6), 1073–1102.

Klein, N. (2007) *The Shock Doctrine*. London: Penguin.

Kloppenburg, J.R. (2004) *First the Seed: The Political Economy of Plant Biotechnology* (2nd edn). Madison, WI: University of Wisconsin Press.

Kubota. (2014) *Annual Report* [online]. Available at: www.kubota-global.net/csr/report_past/pdf/2014/14alldata.pdf (accessed 22 January 2016).

La Vía Campesina. (2011) *What is La Vía Campesina?* [online]. Available at: http://viacampesina.org/en/ (accessed 22 January 2016).

Mamdani, M. (1996) *Citizen and Subject: Contemporary Africa and the Legacy of Late Colonialism*. London: Currey.

Mayozer, M. and Roudart, L. (2006) *A History of World Agriculture: From the Neolithic to the Current Crisis*. London: Earthscan.

McCann, J. (2005) *Maize and Grace: Africa's Encounter with a New World Crop, 1500–2000*. Cambridge, MA: Harvard University Press.

McMichael, P. (2008a) *Development and Social Change: A Global Perspective*. London: Pine Forge.

McMichael, P. (2008b) Peasants Make Their Own History, But Not Just as They Please... In: Borras, S.M. Jr., Edelman, M., and Kay, C. (eds), *Transnational Agrarian Movements Confronting Globalization*. Oxford: Wiley-Blackwell, pp. 37–60.

McMichael, P. (2010) Food Sovereignty in Movement: Addressing the Triple Crisis. In: Wittman, H., Desmarais, A.A., and Wiebe, N. (eds), *Food Sovereignty: Reconnecting Food, Nature & Community*. Oxford: Pambazuka, pp. 168–185.

McMichael, P. (2012) The Land Grab and Corporate Food Regime Restructuring. *Journal of Peasant Studies*, 39(3–4), 681–701.

Meiksins-Wood, E. (2002) *The Origins of Capitalism*. London: Verso.

Meredith, M. (2005) *The State of Africa*. London: Simon & Schuster.

Millstone, E. and Lang, T. (2008) *The Atlas of Food*. Brighton: Earthscan.

Mitchell, D. (2013) 'The Issue is Basically One of Race': Braceros, the Labor Process, and the Making of the Agro-Industrial Landscape of Mid-Twentieth-Century California. In: Slocum, R. and Saldanha, A. (eds), *Geographies of Race and Food: Fields, Bodies, Markets*. Farnham: Ashgate, pp. 79–96.

Moyo, S. and Yeros, P. (2005a) *Reclaiming the Land: The Resurgence of Rural Movements in Africa, Asia and Latin America*. London: Zed Books.

Moyo, S. and Yeros, P. (2005b) Land Occupations and Land Reform in Zimbabwe: Towards the National Democratic Revolution. In: Moyo, S. and Yeros, P. (eds), *Reclaiming the Land: The Resurgence of Rural Movements in Africa, Asia and Latin America*. London: Zed Books, pp. 165–205.

Newell, P. (2008) Trade and Biotechnology in Latin America: Democratization, Contestation and the Politics of Mobilization. In: Borras, S.M. Jr., Edelman, M., and Kay, C. (eds), *Transnational Agrarian Movements Confronting Globalization*. Oxford: Wiley-Blackwell, pp. 177–208.

New York Times. (2015) Room for Debate: The Future of Trans-Pacific Trade [online]. *New York Times*. 6 October 2015. Available at: www.nytimes.com/roomfordebate/2015/10/06/the-future-of-trans-pacific-trade (accessed 22 January 2016).

Nierenberg, D. (2005) *Happier Meals: Rethinking the Global Meat Industry*. Washington, DC: Worldwatch.

Oxfam. (2008) *Notes on Inequality and Poverty in Brazil: Current Situation and Challenges* [online]. Available at: www.oxfam.org.uk/resources/downloads/

FP2P/FP2P_Brazil_Inequality_Poverty_BP_ENGLISH.pdf (accessed 22 January 2016).

Patel, R. (2007) *Stuffed and Starved: Markets, Power and the Hidden Battle for the World Food System*. London: Portobello.

Pimple, M. and Sethi, M. (2005) Occupations of Land in India: Experiences and Challenges. In: Moyo, S. and Yeros, P. (eds), *Reclaiming the Land: The Resurgence of Rural Movements in Africa, Asia and Latin America*. London: Zed Books, pp. 235–256.

Pretty, J. (2009) Can Ecological Agriculture Feed Nine Billion People? *Monthly Review*, 61(6), 46–58.

Reuters. (2015). Hundreds of Thousands Protest in Berlin Against EU–U.S. Trade Deal [online]. *Reuters*. 10 October 2015. Available at: http://nyti.ms/1VLiqUr (accessed 22 January 2016).

Revkin, A.C. (2014). Forget the World Cup – Brazil Posts Double Win with Simultaneous Soy Boom and Deforestation Drop [online]. *New York Times*. 5 June 2014. Available at: http://nyti.ms/1tM3fqI (accessed 22 January 2016).

Sage, C. (2012) *Environment and Food*. Abingdon: Routledge.

Simmons, C.S., Walker, R.T., Arima, E.Y., Aldrich, S.P., and Caldas, M.M. (2007) The Amazon Land War in the South of Para. *Annals of the Association of American Geographers*, 97(3), 567–592.

Spiegel. (2008) Global Food Crisis: The Struggle to Satisfy China and India's Hunger [online]. *Spiegel*. 28 April 2014. Available at: www.spiegel.de/international/world/global-food-crisis-the-struggle-to-satisfy-china-and-india-s-hunger-a-550943-2.html (accessed 22 January 2016)

Stiglitz, J. (2002) *Globalization and Its Discontents*. London: Penguin.

UN (United Nations). (2010) *Report Submitted by the Special Rapporteur on the Right to Food, Olivier De Schutter* [online]. Available at: www2.ohchr.org/english/issues/food/docs/A-HRC-16-49.pdf (accessed 22 January 2016).

USDA (United States Department of Agriculture). (2014) Food Expenditures: Percent of Household Final Consumption Expenditures Spent on Food, Alcoholic Beverages, and Tobacco that were Consumed at Home, by Selected Countries, 2012 [online]. Available at: www.ers.usda.gov/datafiles/Food_Expenditures/Expenditures_on_food_and_alcoholic_beverages_that_were_consumed_at_home_by_selected_countries/table97_2012.xlsx (accessed 22 January 2016).

USDA (United States Department of Agriculture). (2015a) *Livestock Slaughter 2014 Summary* [online]. Available at: http://usda.mannlib.cornell.edu/usda/current/LiveSlauSu/LiveSlauSu-04-27-2015.pdf (accessed 22 January 2016).

USDA (United States Department of Agriculture). (2015b) *Poultry Slaughter Annual Summary* [online]. Available at: http://usda.mannlib.cornell.edu/usda/current/PoulSlauSu/PoulSlauSu-02-25-2015.pdf (accessed 22 January 2016).

Veltmeyer, H. (2005) The Dynamics of Land Occupations in Latin America. In: Moyo, S. and Yeros, P. (eds), *Reclaiming the Land: The Resurgence of Rural Movements in Africa, Asia and Latin America*. London: Zed Books, pp. 285–316.

Vermeulen, S. and Cotula, L. (2010) Over the Heads of Local People: Consultation, Consent, and Recompense in Large-Scale Land Deals for Biofuels Projects in Africa. *Journal of Peasant Studies*, 37(4): 899–916.

Walker, R.A. (2004) *The Conquest of Bread: 150 Years of Agribusiness in California*. London: The New Press.

Weis, T. (2007) *The Global Food Economy: The Battle for the Future of Farming*. London: Zed Books.

Weis, T. (2013) *The Ecological Hoofprint: The Global Burden of Industrial Livestock*. London: Zed Books.

Wittman, H., Desmarais, A.A., and Wiebe, N. (eds) (2010) *Food Sovereignty: Reconnecting Food, Nature & Community*. Oxford: Pambazuka.

Wolford, W. (2003) Producing Community: The MST and Land Reform Settlements in Brazil. *Journal of Agrarian Change*, 3(4), 500–520.

Wolford, W. (2004) This Land is Ours Now: Spatial Imaginaries and the Struggle for Land in Brazil. *Annals of the Association of American Geographers*, 94(2), 409–424.

World Bank. (2007) *World Development Report 2008: Agriculture for Development*. Washington, DC: World Bank.

World Bank. (2010) *World Development Indicators* [online]. Available at: http://data.worldbank.org/sites/default/files/frontmatter.pdf (accessed 22 January 2016).

World Bank. (2014a) Cereal Yield (kg per Hectare) [online]. Available at: http://data.worldbank.org/indicator/AG.YLD.CREL.KG (accessed 22 January 2016).

World Bank. (2014b) *World Development Report* [online]. Washington, DC: World Bank. Available at: http://siteresources.worldbank.org/EXTNWDR2013/Resources/8258024-1352909193861/8936935-1356011448215/8986901-1380046989056/WDR-2014_Complete_Report.pdf (accessed 13 April 2016).

Young, I.M. (1990) *Justice and the Politics of Difference*. Princeton, NJ: Princeton University Press.

Zoomers, A. (2010) Globalisation and the Foreignisation of Space: Seven Processes Driving the Current Global Land Grab. *Journal of Peasant Studies*, 37(2), 429–447.

3 Food processing

3.1 Introduction

What I tried to do in Chapter 2 was provide an account of the way food production occurs in the heartlands of northern agriculture and beyond. I discussed the input-dependence of northern agriculture, some of its most problematic features, and how it meets up with peasant-based, smallholder agriculture in the spaces of Amin's nightmare. That second stage in the life of food is obviously and emphatically important, hence paying attention to it – and attention to the sorts of foodscapes we find there – sheds light on some striking features about the world we live in today.

In this chapter, I move on from agriculture to consider what happens in the third stage in the life of food: when food processing takes place. At issue is a similar set of dynamics to what we have come across so far. There is the corporate food regime; large and powerful transnational firms; and then a range of odd developments emerging from this general scene. The food processing industry has come to occupy such an important place in the diets of billions of people. So many of us know the products, tastes, and smells of processed food – from breakfast cereals, to canned food and ready-to-eat meals – that we might struggle to consider living without it all. As such, it is important that we grasp some of the key drivers and outcomes of the industry's central place in the overall food economy.

Section 3.2 offers an analysis of food processing and the role it has played in answering the proletarian food question. I focus on two ways the industry has sought to drive down production costs: organizational restructuring and food chemistry innovation. I then use the case study in Section 3.3 to discuss how these innovations have enabled the industry to pursue the production of energy-dense foodstuffs that seek out a consumer 'bliss point' (Moss, 2013a, 2013b) to maximize sales, market share, and profit. In Section 3.4 I look at how this stage in the life of food is oppressive, which I argue is about the way food processors seek to marginalize consumers, for example by pushing us out of the kitchen. I argue that, in their systematic effort to construct a food economy that suits its interests over the interests of society at large, these sorts of firms are structurally at fault and guilty of marginalizing other processes and practices. Finally, I use Section 3.5 to shine a light on some of the more promising developments to emerge in recent years, particularly the creation of new niche products that might slowly bring about a less socially destructive food processing industry.

3.2 The world of food processing

Humans have been taking basic foodstuffs and 'processing' them into something else for a long time. Take bread, as one of the most obvious examples (on this, see Bobrow-Strain, 2013). Bread needs to be processed, made; it is about turning its constituent ingredients – water, sugar, salt, flour, and yeast – into something else. It takes skill, knowledge, and investment. It can go wrong. But we have been making bread for a long time; developing our competency, finding new ways to mix different types of grains with other ingredients at our disposal. In effect, then, we have been *processing* these ingredients and making an entirely new, less perishable, and more mobile foodstuff. The core element of food processing – transformation to improve the use value of all the constituent ingredients – is therefore something quite personal and familiar to many of us.

Although bread-making was always something that happened in the home, small enterprises – pre-capitalist, capitalist, and non-capitalist, such as cooperatives – gradually became involved in bread-making (and other forms of processing food). The humble local bakery might stand out in our imagination here: a place where bread and cakes are made fresh on a daily basis, freeing up the buyer from making their own bread, delivering a much-needed and much-loved product. These sorts of bakeries make bread on a larger scale than the individual household and they inevitably find ways to speed things up or bring down costs (see Lawrence, 2004; Sage, 2012: 165). They are part of a larger food processing industry, which uses the selection of ingredients in novel ways; finds innovations; and drives down costs. Thus, far from the ancient household and the small bakery, we have now reached a stage where bread is baked in giant bakeries run by capitalist firms. In the US, this transition occurred between 1890 and 1930: initially bread was made in 'one-oven shops with three or four employees [but] by the late 1920s, large bakeries regularly churned out 100,000 loaves a day' (Bobrow-Strain, 2013: 266).

This sort of shift has occurred alongside the development of a wide range of other food processing operations that nowadays prepare the meat going into our burgers, make and package the pre-cooked meals thrown into our microwaves at home, and lots more besides. In between the providers of agricultural inputs and the farmer at one end of the food system, and then the retailers and consumers at the other end, there is a vast industry constituted by a wide range of food processing firms. These businesses draw upon changes in agriculture to churn out vast quantities of processed food. In so doing, they have driven down the price of food. In developing the industry and the skills and knowledge that go into making its diverse range of products, the food processors have gone a long way toward answering the proletarian food question. For example, the proportion of total household spending on food in the US has fallen from around 42.5 per cent in 1900 to 13.1 per cent in 2002–2003 (US Department of Labor, 2006). In the UK, meanwhile, 11.6 per cent of household income was spent on food and non-alcoholic drink in 2012 (Office of National Statistics, 2013), whereas in the late 1950s households spent around one-third (Carolan, 2011: 2).[1] Buying food has become less of a financial burden. Partly at issue for us, I think, is trying to understand how this happened. In one sense, of course, it is about the sorts of developments I reviewed in Chapters 1 and 2: the innovations upstream of the farm and then the growth of northern agriculture. But the food processing industry has also managed to drive down production costs and lower the eventual price consumers pay to eat. I think we can explain this in two broad ways. Let me deal with each of these in turn.

3.2.1 Organizational restructuring

What I mean by 'organizational restructuring' is that food processors, like other capitalist firms with vast investments, have paid close attention to how they might economize by adjusting relations with other firms and with the workers and suppliers they rely upon. For instance, a prominent feature of the food processing industry's capacity to cut costs and expand output has been the drive to consolidate via purchases of other smaller firms, or if not via acquisition then at least via horizontal strategic alliances (e.g. on the case of alliances between SPC Ardmona, an Australian firm, and Siam Foods in Thailand and Rhodes Food Group in South Africa, see Hattersley *et al.*, 2013). The outcome of all this buying and merging is the formation of food giants with a global reach. According to data on the food and beverage sector compiled by the United Nations Conference on Trade and Development (UNCTAD), in 2007 the largest food firm was Nestlé, which had global sales of $95 billion and employed 276 000 people worldwide (UNCTAD, 2009: 242). Kraft (makers of well-known brands such as Dairylea and Philadelphia cheese) and Unilever (whose brands include Hellman's Mayonnaise), the next two largest food firms, had $37 billion and $59 billion in global sales. Figure 3.1 presents some more up-to-date data on some well-known food processing firms.

All of these giant firms – and many of the smaller ones, too – have a long history of acquisitions. To use Kraft as an example, it was bought by the cigarette company Philip Morris in 1988, which then bought RJR Nabisco in 2000 and combined the two food giants into one firm, Kraft Foods Inc., which continued the consolidation process by purchasing Cadbury in 2010. But Kraft Foods Inc. demerged in 2012, splitting off into two firms: the first kept the Kraft name, while the other became Mondelēz. It is precisely

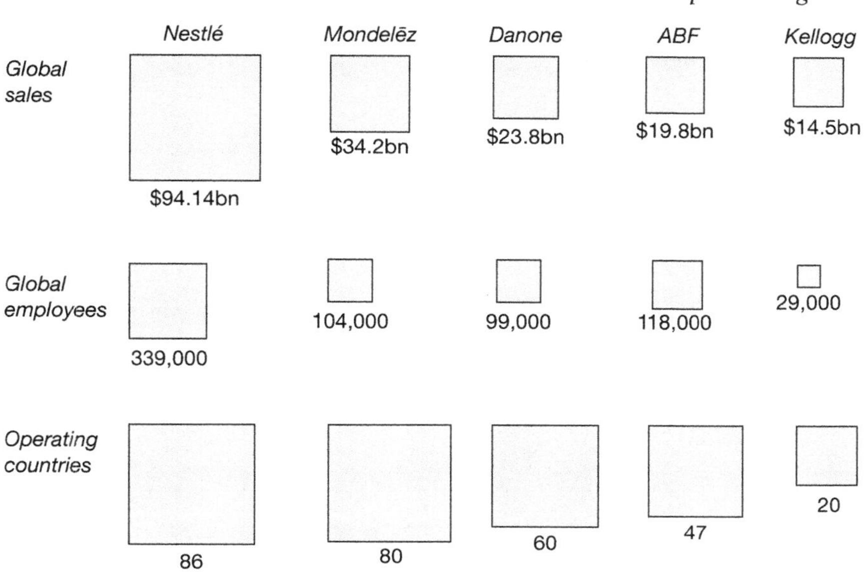

Figure 3.1 Global sales, number of employees, and operating countries of five food processing firms, 2014 (sources: ABF (2014), Danone (2014), Mondelēz (2014), Kellogg (2014), Nestlé (2014)).

through these mergers and demergers, acquisitions and then splits, that the food industry changes, giving it that highly inconstant feel, even if the brand names stay the same. And the practice is not just pursued by the giants: Irish firm Kerry Group, for instance, with a revenue of almost $8 billion in 2013, has grown in part by acquiring firms such as York Dragee in 2000, Voyager Foods in 2001, EBI Foods in 2002, Quest Food Ingredients in 2004, Headland Foods in 2011, and many others (Kerry Group, 2014).

Acquiring other firms increases the likelihood that food processors will consolidate activities and ultimately reduce costs. For example, if Unilever owns a factory where hot dogs are produced and it buys a rival firm that also produces hot dogs, it can consider closing the rival firm's plant, laying off some or all of its workers, and expanding production at its own plant (or vice versa, if the rival firm's plant is more efficient). Alternatively, cost reductions might be achieved simply by virtue of the newer firm's greater buying power, its capability to buy in bulk, even to force suppliers – of raw materials from farms or other processed materials, such as flavourings or preservatives – to offer lower prices. Super-sizing brings with it power, which in this case can be used to drive down costs and increase profitability. And because the food sector as a whole is still overwhelmingly constituted by small firms (e.g. see Sage, 2012: 167), there remains plenty of scope for further consolidation.

Figure 3.2 A food processing factory, China (source: chinahbzyg/Shutterstock).

It is helpful to view this process of consolidation as a reaction to the intense pressure exerted on food producers by *their* customers, the supermarkets. We will consider the role of supermarkets in the contemporary global food economy in more detail in Chapter 4. For now, what is important to note is that the supermarkets demand that food producers reduce costs, diversify, and fill their shelves with affordable product. Supermarkets can play one supplier off against another, perhaps by offering them a better position on their shelves or even in some circumstances refusing to stock items if suppliers do not agree to a low enough price. In addition, supermarkets produce 'own brand' processed food items. The result for many food processing companies is intense competition to produce the right type of product at the right price. In turn, food processors – already experienced cost-cutters – have looked for ways to deal with the pressure.

One avenue is to re-arrange labour processes, such as by intensifying work practices, including the introduction of round-the-clock production and new shift patterns, which undoubtedly increase the physical and mental strain on workers (see Neupane *et al.*, 2014). Firms may even try to cut corners with respect to safety (perhaps only until regulators such as the Occupational Safety and Health Administration step in, as has happened recently with ConAgra in the US (Occupational Safety and Health Administration, 2014)). And in many locations around the world, such as Indonesia, firms as big and profitable as Nestlé have refused to recognize trade unions (IUF, 2012). There is, however, and probably as we should expect, a fair bit of unevenness here. In France, for example, national minimum wage laws and the extension of collective wage agreements to all workers in meat processing and confectionery have meant that working conditions are not as bad as they might be, or as retail chains might want (Caroli *et al.*, 2009). Partly as a result of that very geographical unevenness in regulatory conditions, some firms have demonstrated their willingness to construct organizational structures that span the globe. Unilever now focuses on producing a relatively small number of globally recognized brands in 150 major sites, 'with a further 200 sites for manufacturing local market-leading or nationally profitable products' (Dicken, 2011: 293). Other large firms such as Nestlé have created macro-regional factories conducting the initial stages of production, with smaller finishing sites that adapt products to local markets (Dicken, 2011: 293, citing Palpacuer and Tozanli, 2008: 82–83).

'Offshoring' is another strategy. Some firms have explored ways of moving labour-intensive parts of their production offshore to lower-wage regions, including by moving production from the US to neighbouring Mexico, sometimes with unfortunate food safety consequences (Strom, 2013). Then there are cases such as the Doux group, a major French and European producer of poultry, which bought Frangosul, 'the fourth largest poultry producer in Brazil, where production costs undercut those in France by two-thirds' (McMichael, 2005: 284). Elsewhere in the meat processing industry, firms have managed to pass on many of the costs to the farmers who rear chickens or pigs on a contract and high-debt basis (Weis, 2007: 79–80). Led by US firm Tyson Foods, which processes meat on an enormous scale – 41 million chickens, 391 000 pigs, and 135 000 head of cattle every week (Tyson Foods, 2014) – the industry has cut the end price consumers pay for meat: whereas in 1930 a whole and dressed chicken sold for $14.27 per kilo in today's money, firms such as Tyson have driven down the production costs so much that it now sells for an average price of $3.45 per kilo (Kristof, 2014). Thus, just as the supermarkets have found ways of exerting pressure on them, the food processors have looked for ways to squeeze *their* suppliers, that is, farmers.

In addition to making poultry farmers subordinate to buying firms such as Tyson, consolidation of the processing sector means farmers have a falling number of potential customers – a tendency towards monopsony – and the danger that they might even become reliant on only one customer, such as a Unilever or Nestlé affiliate. In this sort of case, there is a strong likelihood that farmers will have to accept a lower price for their commodities. Alternatively, as processors require bulk quantities of crops, their growing size puts pressure on farmers to increase the scale of their operations by buying up neighbours or by adopting unsustainable technologies and practices (on the often-strained relations between farmers and food processors, see Clapp, 2012: 114–116; also Carolan, 2012: 201–202; Sage, 2012: 37–38).

A final noteworthy issue here is that, in their drive to push down product costs, processors pressure farmers to grow the types of crops that processors are accustomed to using or crops that travel well to where they are processed. One result is 'genetic erosion', a feature of the food system that worries many experts because '90 per cent of the world's food supply comes from only fifteen species of crop plant and eight species of livestock' (Sage, 2012: 100, citing Pimentel and Pimentel, 2008). Then there is the loss of crop diversity in specific places, such as in China where 'nearly 10,000 wheat varieties were cultivated in 1949 [but by] the 1970s, only about 1,000 varieties were in use', while in Mexico, only '20% of the maize varieties reported in 1930 are now known' (Shand, 1997: 22). Crop genetic diversity is highly valuable to the long-term sustainability of the overall food system, not least because it helps farmers minimize risk and manage uncertainty, yet the sort of lock-in food processors require in the short term is a major threat to that diversity. If the lower final price for processed food is an outcome we might all cheer about, it comes with some hidden built-in prices.

3.2.2 Food chemistry innovation

Like all capitalist firms in pursuit of profit, all of these food processors have to keep a firm eye on innovation, looking at the foods they process, finding new ways to do things. As I have noted, the food processing industry has found ways to cut costs and expand output by consolidating and squeezing its suppliers, but it has also focused on developing its understanding of how to manipulate the properties of food. We can see some outcomes of this experimentation around us (microwave popcorn, bagged salad, all of the various new cooking sauces, frequently given new packaging to catch our eye); but there are many other unseen, obscure but still significant innovations to consider. Fundamentally at issue here is that the food processing industry has developed a sophisticated level of knowledge about how to truly put food chemistry to work.

Up until about 50 years ago, the industry relied on relatively basic techniques. The main task was slowing down the pace of food decay, for example via 'drying, reducing ambient temperature, or employing one of a number of preservation techniques such as salting, smoking or pickling' (Sage, 2012: 162). None of this was all that different from what humans had been doing on a smaller scale for many years: smoking fish, salting meat, or pickling vegetables to ensure they last longer. But the food processing industry has found new ways to do this, for example via the application of preservatives such as calcium propionate (labelled E282 on food packaging), potassium nitrate (E252), or sodium nitrite (E251) (Russell and Gould, 2003). These additives reduce the growth of fungi or bacteria and thereby keep products looking and smelling fresh, even if they were made months or even years ago. Other techniques include vacuum packing, which sucks the air out and prevents the growth of bacteria; irradiation, which treats foods with high-energy electrons or X-rays to kill bacteria, mould, and insect pests; and just the plain old use of freezing food (e.g. on frozen food from

California see Walker, 2004: 22). All of these interventions change the properties of food; they also change what consumers expect food to be like, how long it should last, and in a sense how disposable it might be. Food processing has gone a long way towards re-making our relationship with food. Consider here how the importance of sales in supermarkets and convenience stores means packaged processed food has to stay on the shelf for quite some time. Once they bring it home, customers expect to be able to hold on to their food without it spoiling.

Another prominent example is 'fractionation', which breaks down foodstuffs into constituent compounds such as proteins, sugars, starches, and alcohols, all of which can then be chemically modified and used as building blocks for other, more complex foods (Sage, 2012: 163). Corn, for instance, can be broken down so much using this process that a wide variety of new, previously unknown commodities can be produced, such as ethanol for fuel, fructose sweeteners to replace cane sugar, and hydrogenated fats for cooking.[2] The industry has sought ways to increase the 'interchangeability between a range of inputs, natural and synthetic' (Sage, 2012: 163). For example, if cane sugar prices are lower than corn, which yields high fructose corn syrup (HFCS), then processors can alter their production processes to use whichever is cheaper, cane or HFCS. The same goes for the interchangeability between maize and soy or other ingredients. What the processors have managed to learn is a range of ways to adjust recipes without altering the final taste, feel, or appearance. Ultimately, what they desire is to be freed of worrying about how seasonal or long-term environmental changes in one region or another might impact on the price they need to pay for raw materials. The technological fix they desire is intended to liberate them, give them buying power and flexibility. In a global market in which agility is so necessary, innovations are about driving down costs but also giving processors scope to react to market changes.

In turn, its knowledge about food chemistry has enabled the industry to develop products that respond to changing societies. For example, in response to the growth of female employment, the industry has developed thousands of processed products targeted at women who remain overwhelmingly burdened with the responsibility of putting food on the table. As one woman in a study led by Engler-Stringer (2010: 217) notes,

> At our house we are a family, so even if I want to go out, even if I am not feeling well, well I still have to go grocery shopping, even if I have a fever of 42 degrees [108 degrees Fahrenheit]. I have to go out. I have to go for food for my son and my boyfriend.

When budgetary constraints and the difficulties of scheduling shopping trips and meals get factored in, processed foods are appealing, especially when they are relatively cheap and easily prepared. Thus, the industry has helped fill but also create a niche for its products, such as pre-cooked meals, sauces, or desserts. Food processors conduct research, learn, experiment, and create products that cater to the changing social circumstances we find ourselves navigating. A contemporary social question is how the time-stressed worker can find the time to eat and get back to work. The food processing industry's answer has been the development of snacks to be eaten on-the-go, including the chocolate bars and crisps which proliferate in convenience stores and vending machines, but even healthier options such as pre-prepared salads sold in plastic bowls with dressing in a sachet and a plastic knife and fork. Underpinned by its knowledge of food chemistry, the industry has an extraordinary capacity to develop products that seek to solve these sorts of social dilemmas.

But if innovations in food chemistry have enabled the industry to expand production, drive down costs, and develop products that respond to societal change, the industry also now has significant scope to invent entirely new foodstuffs – indeed to do almost magical things with food. A case in point here is 'ultra-processed' food, which includes products such as 'burgers, frozen pasta, pizza and pasta dishes, nuggets and sticks, crisps, biscuits, confectionery, cereal bars, carbonated and other sugared drinks, and various snack products' (Moodie *et al.*, 2013, in Monteiro *et al.*, 2013: 22). As I expect we all know, because no doubt we have been lured into buying these sorts of products at some point in our lives (if not on a daily basis), they are made to appear attractive, with unnatural bright colours and novel textures, and always skilfully packaged and advertised. They tend to be 'hyper-palatable', that is, easy to eat, munch, or swallow in one; ready to consume at any time with no need for baking in the oven; energy-dense, with high fat and sugar content and low nutritious value; cheap to buy but highly profitable for the producer.

Again, these sorts of products emerge from careful studies of what consumers want. In effect, this is about food *engineering*, not just food chemistry. It is about manipulating the properties of proteins and carbohydrates, transforming everything, but in a controlled way, in a stable way that serves a specific purpose: sales revenue growth, customer loyalty, market share, and profit. It should go without saying here that much of this engineering of ultra-processed foods results in extremely unhealthy products, especially if consumed frequently and in large quantities, which the producers want consumers to do. But profits trump human health. Consider here the industry's continued use of 'trans fats'. Trans fats stem from partially hydrogenated oils – formed when liquid oil is treated with hydrogen gas and made solid – and are cheaper to use than saturated animal fats like butter. They raise the level of bad cholesterol and lower levels of good cholesterol and are therefore a major contributor to heart disease. But food processors like them because trans fats are cheaper to use than other oils and because they have unique properties, such as giving a nice flakiness to baked goods. Although public health campaigns at urban, state and national level have forced producers to reduce their use in recent years, trans fats can still be found in cake icing, microwave popcorn, packaged pies, frozen pizzas, and some margarines. The industry knows trans fats should be scrapped, but still they are used (see Tavernise, 2013).

Sugar is another case in point. That search for the 'bliss point' is underpinned by 'cardinal rules' in the industry, such as 'When in doubt, add sugar' (Moss, 2013b). One danger of just adding sugar is that the human body cannot easily absorb it, hence the risk that eating too much sugar-rich ultra-processed food will result in health problems, particularly obesity and type 2 diabetes. Even so, the industry churns out vast quantities of sugar-rich products, ranging from sweetened dairy products and breakfast cereals, to snacks and various types of soft drinks. Scientific understanding is clear on this issue: too much sugar is bad for the human body (American Heart Association, 2009). Still the industry persists. It knows customers have a taste for sugar and that any recipe for sales and profits must give serious consideration to pumping in unhealthy amounts of it. In its attempt to defend its interests, the food industry argues that individuals can choose not to eat sweet foods or drinks, that consumers are free and responsible. But this seems hard to accept, given how much ultra-processed food the industry produces and how it pushes it onto the market at a price well below more nutritious food. In an industry where 'just add sugar' is a cardinal rule, the obvious conclusion to draw is that the engineering of ultra-processed foods illustrates how little the industry really cares about the effects its products might have on consumers.

All of this discussion points to a broader conclusion about the food processing industry: the issue is not just that it produces ultra-processed foods, but rather that the development and proliferation of ultra-processed foodstuffs reflects a dominant mentality and vision of its customers. The firms adopt certain imaginaries about us, the consumer, of our bodies, of what we might and even should be able to bear. The industry sees consumers as mere conduits for delivering their products into our guts. Our role as consumers is to buy and consume their products, even if the outcome is bad for us. We are studied, observed, tested; understood, theorized, imagined. Our bodies are constituted by distinct areas or spaces that the processors try to conceptualize and then develop products to suit: there is the mouth and its 'mouth feel' and then the stomach and its 'stomach space'. What the industry is engaged in is very much a struggle to find the best way to blend ingredients such that products can achieve the best mouth feel – the degree of chewiness, say, or the right amount of crunch – and taste and enjoyment that a 'bliss point' (Moss, 2013a, 2013b) will be reached and consumers will want to buy more of the product, thereby increasing the amount of stomach space taken up by a KitKat or new line extensions such as the chocolate versions of Kraft's Philadelphia Cream Cheese, or whatever the next big product hit will be. And the whole process – repeated over and over again with new and newer product lines, new extensions of existing products, and re-invented packaging styles – is intended to become so expected by consumers that the power of the industry is normalized, its place in the making of the contemporary global food economy secured.

It should not be too surprising that this way of using food chemistry has made ultra-processed food increasingly dominant globally, that is, in the richer countries where the food giants first developed knowledge about making and selling these products *and* in new so-called 'emerging economies' such as Brazil, Thailand, Mexico, and China. In part, the change in places such as Brazil or Thailand is similar to changes occurring in the US or Western Europe: more formal employment, more women working outside of the home, and rising incomes among some middle-class people meaning disposable incomes can be spent on treats such as snacks, or in fast-food restaurants where ultra-processed food pervades. But it is also important to note how these sorts of countries have been *targeted* by the food giants, even to the point where a new mantra has developed such as how these firms will aim to 'teach the world to snack' (Monteiro *et al.*, 2010) – which means snacking on the ultra-processed, energy-dense, sweet and fatty products the food giants churn out and advertise. Snacking between meals in Chile or Indonesia might once have simply entailed eating some fruit; now, the food industry targets consumers with advertisements that emphasize the nutritional value and enjoyment of snacking on chocolates, crisps, and cakes. No wonder these snacks add up to one-quarter of all calories consumed in many of these 'emerging' economies (Monteiro *et al.*, 2010). Of course, for the food giants, these sorts of developments are ideal; exactly what they need to overcome static or even declining sales. What better way to address falling sales in the US, say, than to develop and open up new markets in Brazil or China? Clearly, the era of neoliberal-style free trade agreements (as I discussed in Chapters 1 and 2) helps food processors. Now almost the whole world is their market, largely unfettered by regulations. Rising sales of ultra-processed food – reflecting advances in food chemistry – have been a huge boon to the food firms and an almost-logical outcome of the corporate food regime.

3.3 Case study: soft drinks and snacks

As I have just discussed, the food processing industry has developed ways of reducing the final relative price consumers pay for food, especially in richer countries. This has been achieved by driving down costs, often via consolidating, sometimes via offshoring, as well as by finding ways to squeeze its suppliers. In effect, a major factor has been the food industry's capacity to configure a complex, resilient but flexible geography. In arranging and connecting diverse sites across the world, these firms have developed large, sometimes immense, manufacturing centres that draw in a diverse range of raw materials, often from the other side of the world, and process them using a mixture of skilled and low-skilled workers. From the research laboratories come recipes, new ideas and ingredients, new ways of processing;

from farms near and far come the raw materials, the cereals, the dairy products, the tropical fruits; from other firms come processed ingredients such as oils, preservatives, and flavourings; and from market research comes knowledge about what to make.

Thus, on production lines all over the world, all this diversity of knowledge and practice and materiality is mobilized, put to work by the food processing firms and eventually shipped out to supermarkets, convenience stores, and restaurants. The outcome of all this mobilization, all this effort to stitch together fixed places and other flows, is increasing industrial capacity, which means more agricultural product can be purchased, churned, baked, twisted, transformed. Moreover, as a dispersed but connected network of producers, the food processing industry is dynamic and inventive. They operate in a competitive market. Under pressure from retailers to offer goods at even lower prices, the industry *must always* keep a firm eye on innovation. That almost goes without saying: these are capitalist firms, hence product or process innovation is always on the cards, given worries about share prices and the threat a competitor might eat into their market share. The numerous and various technological developments brought together in the food processing industry to secure cost savings reflect this peculiar pressure to survive that capitalist firms are subject to. But yet, if this pressure to innovate is partly about survival and, from the perspective of the capitalist firm, defensible, some practices in the industry are nevertheless quite controversial. To put it bluntly, innovation in food processing has generated a vast range of products, many of which are unnecessarily bad for us. Soft drinks, otherwise known as sugar-sweetened beverages, are a prime example here.

Most public health experts agree that we should restrict our consumption of soft drinks, especially if we want to avoid gaining weight. A 330 ml can of Coke, for example, contains around 35 g of sugar, which is about as much sugar as the American Heart Association (2009) recommends adult men consume in a day. But the market for soft drinks is huge, with billions of dollars to be made. In 2012 alone, Coca-Cola made around $11 billion profit from $48 billion sales (Nestle, 2015: 92). Given what is at stake, the large drinks firms must invest in finding ways to keep and expand the number of their customers. They might create new designs on their bottles and cans, or develop new advertisements and sponsorship partnerships with sports or television shows. However, another important innovation strategy is to invest in what the industry refers to as 'line extensions', which take the basic concept of their product and re-imagine how it might be combined with new ingredients (see Carolan, 2011: 69–70).

In his recent book *Salt, Sugar, Fat: How the Food Giants Hooked Us*, Michael Moss (2013b) discusses this process of line extension. Cadbury Schweppes wanted to create a new version of Dr Pepper, one that would allow the company to occupy more shelf space in supermarkets and convenience stores. After conducting intensive research, the firm found a product it believed would work. The magic ingredients were cherry and vanilla flavourings, which were mixed with their regular Dr Pepper syrup. They varied the amounts of each ingredient and tested them on consumers to see which blend gave the best response. Ultimately, they were searching for the 'bliss point': the specific blend of ingredients which most satisfies consumers while minimizing production costs and therefore maximizing profits. A report resulting from Dr Pepper's research included details on aspects such as how respondents felt about 'mouth feel' – 'the way a product interacts with the mouth, as defined more specifically by a host of related sensations, from dryness to gumminess to moisture release' (Moss, 2013b) – colour, and flavouring. It concluded that a new product, Cherry Vanilla Dr Pepper, satisfied consumers and that at a certain point the firm could reduce its use of their Dr Pepper syrup without affecting consumer 'bliss'.

It is via studies such as this that food firms make the sorts of minor tweaks and adjustments to reduce production costs and boost profits in many of the other processed products we come across on a daily basis. The new flavours of soups, the new packaging of breakfast cereals, or the line extensions of existing products: these appear on the supermarket shelf based on extensive research and investment, including research that has as its basis the aim of maximizing consumer bliss, even at the expense of consumer health. In this regard, consider how firms develop snacks such as crisps.

Crisps are in many senses the perfect junk food. They contain sugar in the form of potato starch; fat because they are fried; and plenty of salt and other, sometimes quite addictive, artificial flavourings. They are engineered carefully, with consideration given to the size of each crisp, the thickness, and strength. Crisps are also relatively durable, mobile, and thanks to various preservatives they can be kept fresh and crisp in their bag for quite some time. Again, as with sweetened soft drinks, the industry knows these products are unhealthy. But competition in this large market is so intense that innovation and invention are necessary to retain market share: if firm A fails to innovate, firm B will probably move on in. Reflecting the inconstant nature of the capitalist food economy – that changeability, that constant appearance of new products intended to capture our taste buds – the market for snacks is always developing, with new products appearing along with numerous new sales pitches.

A case in point here is the development of Ruffles potato chips by Frito-Lay, owned by PepsiCo (see *Advertising Age*, 2013). Frito-Lay noticed that consumers were increasingly using crisps to grab dips – blue cheese or barbecue, usually manufactured in bulk by the processors, packaged in attractive cans, and sold next to crisps in the supermarkets. Researchers found that some crisps were not strong or thick enough to pick up the dip. What Frito-Lay struck upon, therefore, was a market for a new, stronger crisp with thicker ridges and new flavours that could match up with the sorts of dips consumers were buying. And, in response, and after studying men's eating habits – watching them eat in bars, or at home, asking them questions in-person as well as online – they developed a new line of Ruffles potato chips, such as Deep Ridged Classic Hot Wing Chips and MAX Beer-Battered Onion Ring Flavored Potato Chips. Both were launched with marketing campaigns and packaging aimed at men. The nutritional content of these new lines speaks volumes about what Frito-Lay thinks men can bear. The Deep Ridged Classic Hot Wing Chips, for example, are sold in a 7.5 oz bag and, as such, contain more than the US recommended daily limit of fat, and just under half of the daily limit of salt and carbohydrates. They are not what men need, but they suit what Frito-Lay need: a palatable, ready-to-consume, energy-dense and in many senses addictive product on the shelf.[3] These cases therefore highlight ingrained practices, learned and developed by the food processing industry and mobilized on their production lines throughout the world. They rely on a certain imaginary of consumers as having rugged bodies, capable of consuming vast amounts of sugar or fat; capable of enduring emerging health problems. It is an insidious way of imagining consumers. And these ultra-processed products are just two of a massive range of products that experiment with ingredients to find a bliss point that achieves high sales at the expense of consumer health.

One health concern, for instance, is that a diet rich in salt and fat, alongside cigarette smoking, raises the likelihood of an individual developing heart disease. Salt is a particular worry because it raises blood pressure, which can trigger strokes. Eating too much ultra-processed food is problematic because it tends to contain high sodium levels as a preservative: shelf life is all-important. Given high rates of heart disease in rich countries, medical knowledge about the best sorts of procedures to pursue has developed enormously, indeed to such an extent that heart disease need not be a killer if the correct treatments and medicine are provided. Rich countries, where rates of heart disease were once the highest in the world, have seen significant reductions in the last two or three decades. Helping achieve these declines have been public health initiatives encouraging people to lower their salt intake. Likewise, some new regulations have been introduced that ask food processors and restaurants to reduce how much salt they add to food. In the UK, for example, the Food Standards Agency followed practices in countries such as Japan and Finland by targeting salt levels in processed food and setting salt-reduction targets over time. As the industry met those targets, consumption fell by 15 per cent from 2001 to 2011, which has helped reduce average blood pressure and resulted in a fall in the number of deaths from heart attacks and strokes (Farley, 2014). It is, I think, a sign of the times – of our neoliberal times – that the British initiative did not compel but rather only *asked* the food industry to act. In part, this signals the weakness of regulators. And even when public health officials ask the industry to act, as has been the case recently in the US, progress is slow, with firms only agreeing to lower salt in some, but definitely not all, food categories.[4]

But voluntary schemes can have *some* success, a point that demonstrates quite nicely a key feature of the contemporary global foodscape; that is, how it is co-produced by dominant tendencies and resistance. There are openings for further pressure to be exerted on the food giants to change their practices. Even if few critics of the food economy expect such pressure to truly change systemic features of the food economy, this does not mean that incremental changes have no value. If regulators cannot force the food industry to change, it is all the more important that they push as hard as possible to emphasize the value of voluntary schemes.

All this having been said, and as we saw with regard to the food industry's move to target lower-income markets – teaching the world to snack – a key point arises here. If it is hard for governments in some of the world's richest countries to take on the food giants, the situation facing poorer countries is another thing entirely. Up against a rising tide of trade in exactly the sorts of ultra-processed, energy-dense foodstuffs that generate the diabetes crisis, it is hard to imagine how governments with small budgets, weak regulatory power, and limited scope to control imports might find ways to take on the corporate food sector.

3.4 In what sense oppressive?

In some respects the food processing industry has to be applauded for finding ways to drive down the cost of what consumers pay for food. It has gone some way toward answering the proletarian food question by operating at scale and constantly innovating. The industry has certainly played its part in ensuring that workers, consumers, you and me, can at least find plenty of food to buy, if we have the money to do so. Yet, as the discussion above makes clear, many of the industry's practices – embedded within a corporate food regime that prioritizes profits before human health – are highly problematic, even oppressive. There is, for instance, the exploitation that necessarily occurs on the production lines; the cultural imperialism pursued by the firms as they promote a certain type of food consumption over extant diets; and then all of the violence enacted on the billions of animals that are slaughtered to satisfy our demand for burgers, lasagnes, and chicken-and-bean-burritos-in-a-bag. However, the particular form of oppression that stands out here is *marginalization* (Young, 1990), and I think it appears in three ways.

First, consider the sorts of product developments pursued by the food industry in recent years. Think about cooking sauces. It is easy to imagine the industry as having our interests at heart here: cooking sauces are there to make our lives easier, after all. And in many ways, this has been a wonderful innovation. I think: 'Now I can cook my chicken in a Tikka Masala sauce, say, or make a Chinese Orange chicken.' Then there are the white sauces for a fish pie or pasta sauces for my bolognese, and many more besides. Without question, there is convenience here. But at the same time, one outcome of all this innovation is that consumers like you and me begin to lose the cooking skills we might once have had or, at least, that our parents had (Pollan, 2013). This is ideal, as far as Unilever or Nestlé are concerned: the more we forget how to cook, the more products we will buy from them. The upshot is that our role as makers of food is under pressure, squeezed by price and convenience, placed on the margins, pushed to the side by the power of the food processing industry and its drive to develop products at all stages of the cooking process. The ideal future, given this process of marginalization, is that we will spend as little time as possible in the kitchen.

Second, in their drive to consolidate and achieve scale, food processing firms marginalize other more traditional ways of making food. Consider that, whereas a relatively broad range of perishable breads were once made locally in almost all neighbourhoods by small, independent bakeries, today the market is increasingly and overwhelmingly dominated by giant bakeries producing massive quantities of longer-lasting bread (but full of preservatives and other largely unnecessary ingredients) (see Bobrow-Strain, 2013). Like the lost skill of making a white sauce, the skill of bread-making is disappearing from households but also communities, as smaller independent bakeries go out of business or are bought up by the larger firms. As the food processing industry grows, so does its power to further marginalize other ideas and practices. And in the process, the industry strives hard to justify pushing aside other practices. In the US, for example, capitalist firms have pitched their bread as safer than home-made bread (Bobrow-Strain, 2013), just as food processors today emphasize the safety benefits of their new packaging concepts over what the butcher down the road offers.

Finally, the food processing industry has sought to marginalize the democratic process. This should not be a surprise. The same can be said about agricultural supplier firms upstream of the farm, as well as supermarkets. More broadly, in a world deeply shaped by neoliberal ideas and practices (e.g. see Brown, 2015), capitalist enterprises have significant scope, on their own or via industry associations, to lobby and influence, as well as co-opt or challenge democratic processes (see also George, 2015). But the reason I think its efforts to marginalize the democratic process really stand out is precisely that the food industry is not producing what we might refer to as optional goods, such as smartphones. Smartphones we can, although perhaps with some inconvenience, live without. The same cannot be said about food. We need to eat. Although many people around the world, even in cities (e.g. see McClintock, 2014), can still produce a lot of the food they need, many more of us are *entirely* reliant on what capitalist enterprises can bring to us. We therefore rely on the food industry preparing products that are safe and nutritious. It therefore seems reasonable that citizens like you and I would consider the state – and the democratic process – as an actor that can, in theory if not in practice, generate policies and regulations that compel food processors to change their ways, should evidence emerge that such change is necessary. In theory, a good government, one that takes public health seriously, might seek to rein in some of the food industry's influence.

At least in principle, a society in which there is a degree of democracy should be able to trump sectional interests such as the food industry. It is entirely conceivable, if not completely desirable, that policies regarding nutritional information on food labels would only take into consideration the advice of independent experts. Instead, via consultations and lobbying, actions by food industry associations – such as the American Beverage Association – and individual firms have consistently looked to ensure that nutritional advice is biased in their favour (Nestle, 2013, 2015). And wherever new challenges emerge, such as proposals to introduce taxes on sugar-rich food and drink products, industry associations and individual firms have been willing to spend millions of dollars and use a wide range of dubious tricks to protect their interests. In the US, for example, the industry has spent at least $70 million between 2009 and 2012 fighting 'soda tax' initiatives (Nestle, 2015: 367).

Internationally, too, associations such as the International Food and Beverage Alliance (IFBA) work to shape the debate about food and protect the interests of their members. The IFBA consults with and comments on World Health Organization recommendations regarding nutrition: in effect, it monitors and observes what the international public health community is doing and will work with national associations to lobby particular governments to defend its interests should the need arise (see also Nestle, 2013: 379–380). Of course, transnational firms and their associations also operate in the context of trade agreements that give firms legal rights to challenge democratic decisions (e.g. see Friel *et al.*, 2016). The corporate food regime makes public health victories over the food industry hard to achieve.

Thus, as I think the above examples illustrate, the food processing industry consistently seeks to marginalize the democratic process, as well as other practices of eating and cooking that threaten its core location and purpose in the contemporary food system. These forms of marginalization are central to its power, strength, and growth. But there is no inevitability here. On the one hand, there are signs that public health arguments about the food industry's practices are creating some better policies. For example, as I mentioned earlier, new voluntary codes have been introduced in some jurisdictions that ask food processors to reduce the salt content of products (Farley, 2014; also Campbell, 2014). Other prominent examples include efforts by national governments, such as Mexico, to tax sugar-rich drinks and snacks (Villegas, 2013). On the other hand, though, democratic processes are not necessarily bound to produce perfect policies. Nutritional advice can go wrong. Consider only that, where nutritional advice was once to look for a low-fat diet to avoid gaining weight, experts have backtracked, in part because some studies found that cutting back on carbohydrates was more important (O'Connor, 2014).

Perhaps the key lesson, then, is that, although better regulation of the food industry is conceivable, it will never be a straightforward process.

Perhaps the food industry could be regulated such that its understanding of producing food at scale, which it has developed over the last few decades, is put to better use. Consider here just how unrealistic it is to expect everyone to eat less processed food, even to the point of avoiding it altogether (e.g. see Pollan, 2013): we simply do not all have the time, the budget, nor even the equipment or space in our kitchens to cook and eat non-processed foodstuffs each and every day. The question is not about how to remove processed food from the human diet, but rather how we might maximize the food industry's potential to create healthier and more nutritious products at scale.

3.5 Resistance: re-thinking processed food

The easiest future for the food industry is to keep doing the same: compete with each other to find the bliss points in soups, pizzas, chocolate bars and candy, crisps, etc. Altering these processes and practices, the task public health officials would like food regulators to take on, will not be easy. And of critical importance in all this is that food firms are facilitated by the logics underpinning the corporate food regime, which limits the capacity of governments to regulate what firms do. In trying to get their products into our stomachs – in finding ways to advertise, market, target, and develop new mechanisms to get their goods out onto the market – the industry has developed a peculiar vocabulary (bliss point, mouth feel, stomach space) and its capacity to continue doing so is propped up by a specific conceptualization of their right to do so.

But there is more to be said about food processors. It is easy to characterize all of these firms as nasty evil-doers. Yet, so long as critics refuse to hold on to a more nuanced understanding of corporate culture and economics, such a characterization will only be of limited use. The fact is, as much as many food corporations *are* intent on finding bliss points in the food they manufacture, their (in)actions open up a small but growing window of opportunity for other firms to develop and sell more nutritious foodstuffs. This opening is certainly what we are beginning to see take shape around us. The market for food is dynamic, inconstant. There are shifts. And underlying some of the newer, more promising developments are decisions by some firms to go against the grain – to resist the temptation to follow the pack – in ways that respond to consumers who are increasingly organizing, protesting, and pressurizing firms to change their practices.

Consider the market for baby food. Without question, one of the hardest aspects of parenting today is navigating the bewildering array of products on the supermarket shelves. The dairy section is no exception, not least regarding the wide selection of yoghurt products. Certainly, many yoghurt products are targeted at adults, but a large chunk is explicitly aimed at children. There are pictures of Disney characters or bright and colourful packaging designed to catch the child's eye (and which are designed as part of broader marketing campaigns that include television advertisements targeted at children) (e.g. see Nestle, 2013: 396–402). Kids love this stuff. But no wonder. Look closely at the nutritional information: many products contain so much sugar or other sweeteners that one small container of yoghurt will contain more sugar than a child should eat in a whole day. According to advice from the American Heart Association (2009), for example, the maximum sugar intake for adults is six to nine teaspoons per day, which translates to about three or four teaspoons, or about 20 grams, for children younger than eight years. Despite this, it is rare to find only 20 g of sugar in yoghurt for kids. Many have at least 30 g.

Against this backdrop, more and more parents have decided to understand these labels, although many do not have the time to read them while inside the supermarket, and others might not even know what to look for. Reflecting the growing sense among parents – and consumers more broadly – that the food processors cannot be trusted, that they do not have the interests of their customers at heart, blogs and tweets and word-of-mouth have spread knowledge of the sugar content of foodstuffs and the link between processed food and type 2 diabetes. As a consequence, more parents now look at the labels and try to calculate whether they should buy particular products for their children and for themselves.

Some firms have taken note. As noted earlier, a crucial aspect of the food processing world is that consumers are studied, examined, tested, and listened to. There is this sort of 'bio-politics' to food processing: this sense that we are mere objects of study, actors for sure, but primarily in the sense that we receive, deliver, and ultimately valorize the investments made in new equipment, machinery, laboratories, marketing, and so on. That there is growing consumer awareness is well known among the food processing firms. But just because consumers are now more interested in healthier eating does not mean firms will respond. Using sugar, salt, and fat to create that bliss point in processed food is an acquired knowledge, one that firms have invested in and understand; and it works. Re-thinking their practices, re-learning, re-imagining how they might deliver more nutritious products: all this entails extra work, more effort, longer-term thinking, more investment, perhaps even short-term losses. Such an effort is risky. A long-term vision *can* be pursued; certainly many firms endure losses for a period of time while they re-tool, develop new competencies, move into new markets. But doing so requires a bravery on the part of owners and managers, and a willingness to stand up to the sort of scrutiny that will emerge from fund managers or stockbrokers whose investments shape the global foodscape in so many ways. So: maybe there is a new market segment, a new and growing niche out there that wants more nutritious food; but can corporations meet the demand, can they actually turn their ships around sufficiently to begin producing the products that can satisfy what consumers want?

At issue in all of this are questions about resistance. Meeting the new demands from consumers for healthier foodstuffs has meant that, within and between firms, some people have had to push an agenda for change and ultimately for new rounds of investment in healthier product categories. One well-known case is that of Derek Yach, who was a leading public health expert in the WHO but was then hired by PepsiCo in 2006 (*Economist*, 2012; Nestle, 2015: 268–276). At Pepsi, Yach tried to change the firm's ideas about what sort of products it should produce, but his tenure there only lasted until 2012; Pepsi's shareholders made it clear that a strategy to produce more healthy foods would only be sustainable if it also delivered profit growth, which by 2012 had not happened.

If Yach's case has received a fair bit of attention, it merely reflects a broader shift within the food industry. The industry has a bad reputation but there are certainly some moves to develop healthier products. A useful example here is that of an Irish dairy firm. Glenisk was founded in 1987 by Jack Cleary, a Co. Offaly farmer who had 14 children and wanted to find a way to create employment for them all. It was just a regular yoghurt producer at first, but in 1995 the firm decided to only produce *organic* yoghurt. Managing Director Vincent Cleary noted that, 'for the first eight years we were on the fringes by going organic' (Newenham, 2012), but the firm gradually found a market for its products. Consumer tastes were changing and the market for organic products was growing, in part because some parents were exploring the possibility of giving their children organic food. Glenisk had yoghurt for kids but 'it wasn't setting the world on fire' until the firm removed the sugar. Responding to their customers – 'we took the time to speak to our customers' (McCall, 2013) – Glenisk explored how to make an unsweetened product for babies. Of course, the new yoghurt had to 'deliver the same taste, texture, shelf life, appearance and mouth feel' (McCall, 2013), hence the need for experimentation, a significant investment in time and resources. The key was to use a blend of pure fruit and concentrated fruit. The result was a success: the sugar-free baby yoghurt was 'launched in mid-2010 and sales increased by over 100 per cent in 30 months' (McCall, 2013), leading the firm to pursue sugar-free yoghurt for the adult market.

These sorts of innovations have won Glenisk various awards and widespread recognition. Along with other changes – it has its own wind turbine and uses electric vehicles – Glenisk has presented itself as a 'green' firm offering a quality and nutritious product. None of this has been easy. Developing relationships with supermarkets is far from straightforward, for example, and Glenisk's customers such as Tesco, Sainsbury's, and Asda have only been won over via consistent hard work. Further, although the firm prides itself on being 'always at the cutting edge of the organic fresh dairy products industry' (*Irish Times*, 2007), pushing the boundaries entails a level of risk that not all firms are prepared to endure. Glenisk is, however, 'an entrepreneurial success story' (Sheridan, 2006), and one that demonstrates the scope that exists for firms to push against the limits of the food processing industry and develop niche products that respond to what consumers want.

This is a form of resistance: refusing to go along with normalized practices, seeking ways to re-think and re-imagine the economy, albeit while remaining firmly bedded to the need to profit and secure market share.

Notably, as a higher-quality product, something stemming from additional research, Glenisk's baby yoghurt isn't cheap. It is hardly what we might refer to as a 'proletarian' foodstuff, in so far as its sales stem from higher-earning, less price-sensitive households who can afford to spend that little bit extra on products like yoghurt. In these ways, therefore, perhaps we cannot make too much of Glenisk's success. But I think there is still a lesson to take from this: the Glenisk story demonstrates that change in the global foodscape can emerge, not just from what critics say, nor from grass-roots pressure, but also from firms. Indeed, as hard as this might be to accept, not least for those who are critical of corporate control over the foodscape, perhaps the best hope we have of seeing a healthier foodscape emerge lies in food processing firms changing their behaviour and practices. Short of a systemic collapse or a revolutionary change in the global economy, it will be firms like Glenisk that end up pushing the boundaries. And, of course, if they achieve enough (that is, if they gain sufficient market share), then maybe the larger firms will follow, if only because they will be under pressure to retain market share. There is nothing inevitable here, no guarantees that positive changes will indeed take shape; but nor can we rule out the possibility that enough firms will change their stance with regard to food processing that some of the more insidious practices will fade away.

What else matters here?

In this chapter I have focused on marginalization rather than exploitation, which certainly could have come up. If I had chosen to look at exploitation, what issues might I have addressed, do you think?

I only touched on the issue of food scandals because I wanted time to consider other issues. If I had looked more closely at food scandals, what sorts of issues do you think I might have raised?

The food industry is not alone in courting controversy. The tobacco industry and then the cosmetics industry and its use of animal testing should come to mind here. In these latter industries, how have the general public and the state tried to exert some influence over what goes on?

You decide

- Is government regulation the best way to ensure the food processing industry develops less insidious practices?
- Is a world without vast enterprises producing at scale – capitalist or not – conceivable?
- By considering how some firms go against normalized practices, I fear I might have irritated the more radical reader who finds it hard to accept

that entrepreneurs have anything positive to add to society! Firms such as Glenisk are in it for the money, of course. But you decide: is an interest in material gain necessarily in conflict with an interest in social justice, equality, or peace?

Suggested reading

Marion Nestle's work is perhaps the best place to start if you want to know more about the food giants and their flaws. See, for example, *Food Politics: How the Food Industry Influences Nutrition and Health* (2013).

Michael Moss' (2013b) *Salt, Sugar, Fat: How the Food Giants Hooked Us* presents an accessible exploration of how the food industry develops its products.

Finally, see Rachel Engler-Stringer's 2010 study, 'The Domestic Foodscapes of Young Low-Income Women in Montreal', of how processed food enters into domestic foodscapes.

Notes

1 Note here that poor households today spend a much greater proportion of their income on food (between 15 and 17 per cent) than wealthier households (as little as 7 per cent) (UK Cabinet Office, 2008: 24). And in many poorer countries today, households spend even greater proportions on food. For example, just under 25 per cent is spent on food in Mexico, around 35 per cent in Vietnam, while food takes up as much as 47 per cent of household income in Pakistan (USDA, 2014).

2 Obviously, this sort of capability has wider implications, for if corn can be used for fuel, say, it can be diverted from the food market; or if it is used to sweeten food or drinks, there are implications for cane sugar producers, including the likelihood that their earnings will fall.

3 More broadly, food companies are well accustomed to targeting specific markets with advertisements and campaigns. Fleming-Milici *et al.* (2013), for example, demonstrate that US food companies disproportionately target Hispanic populations. Similar research calls attention to campaigns that disproportionately target black populations (Harris *et al.*, 2015).

4 Also in the UK, a recent move to have firms agree to cut calories in snacks has only managed to gain support from a few of the big producers, such as Nestlé, Unilever, and Coca-Cola; other firms, including Danone, Tate & Lyle, and Kellogg's are still refusing to take part (Campbell, 2013, 2014).

References

ABF. (2014) *Annual Report and Accounts 2014: Value Together* [online]. Available at: www.abf.co.uk/documents/pdfs/2014/2014_abf_annual_report_and_accounts.pdf (accessed 13 April 2016).

Advertising Age. (2013) *How the Women Running Ruffles Unscrambled the 'Bro Code'* [online]. *Advertising Age*. 2 June 2013. Available at: http://adage.com/article/news/women-running-ruffles-unscrambled-bro-code/241821 (accessed 22 January 2016).

American Heart Association. (2009) *Dietary Sugars Intake and Cardiovascular Health: A Scientific Statement from the American Heart Association* [online]. Available at: http://circ.ahajournals.org/cgi/reprint/CIRCULATIONAHA.109.192627 (accessed 22 January 2016).

Bobrow-Strain, A. (2013) White Bread Biopolitics: Purity, Health, and the Triumph of Industrial Baking. In: Slocum, R. and Saldanha, A. (eds), *Geographies of Race and Food: Fields, Bodies, Markets*. Farnham: Ashgate, pp. 265–290.

Brown, W. (2015) *Undoing the Demos*. Brooklyn, NY: Zone.

Campbell, D. (2013) Food Packaging 'Traffic Lights' to Signal Healthy Choices on Salt, Fat and Sugar [online]. *Guardian*. 19 June 2013. Available at: www.theguardian.com/society/2013/jun/19/traffic-light-health-labels-food (accessed 22 January 2016).

Campbell, D. (2014) Study Suggests Link Between Fall in Salt Intake and Drop in Heart Attack Deaths [online]. *Guardian*. 14 April 2014. Available at: www.theguardian.com/society/2014/apr/14/research-finds-link-between-drop-in-salt-consumption-and-fall-in-heart-attack-deaths (accessed 22 January 2016).

Carolan, M. (2011) *The Real Cost of Cheap Food*. Abingdon: Earthscan.

Carolan, M. (2012) *The Sociology of Food and Agriculture*. Abingdon: Earthscan.

Caroli, E., Gautie, J., and Lamanthe, A. (2009) The French Food-Processing Model: High Relative Wages and High Work Intensity. *International Labour Review*, 148(4), 375–394.

Clapp, J. (2012) *Food*. Cambridge: Polity.

Danone. (2014) *Registration Document: Annual Financial Report 2014*. Available at: www.danone.com/uploads/tx_bidanonepublications/DANONE_2014_Registration_Document_ENG.pdf (accessed 13 April 2016).

Dicken, P. (2011) *Global Shift* (6th edn). London: SAGE.

The Economist. (2012) Food for Thought [online]. *The Economist*. 15 December 2012. Available at: www.economist.com/news/special-report/21568064-food-companies-play-ambivalent-part-fight-against-flab-food-thought (accessed 22 January 2016).

Engler-Stringer, R. (2010) The Domestic Foodscapes of Young Low-Income Women in Montreal: Cooking Practices in the Context of an Increasingly Processed Food Supply. *Health Education & Behavior*, 37(2), 211–226.

Farley, T.A. (2014) The Public Health Crisis Hiding in Our Food [online]. *New York Times*, 20 April 2014. Available at: www.nytimes.com/2014/04/21/opinion/the-public-health-crisis-hiding-in-our-food.html (accessed 22 January 2016).

Fleming-Milici, F., Harris, J.L., Sarda, V., and Schwartz, M.B. (2013) Amount of Hispanic Youth Exposure to Food and Beverage Advertising on Spanish- and English-Language Television. *JAMA Pediatrics*, 167(8), 723–730. doi:10.1001/jamapediatrics.2013.137.

Friel, S., Ponnamperuma, S., Schram, A., *et al.* (2016) Shaping the Discourse: What Has the Food Industry Been Lobbying for in the Trans Pacific Partnership Trade Agreement and What are the Implications for Dietary Health? *Critical Public Health*, doi: 10.1080/09581596.2016.1139689.

George, S. (2015) *Shadow Sovereigns: How Global Corporations are Seizing Power*. Cambridge: Polity.

Harris, J.L., Shehan, C., and Gross, R. (2015) *Food Advertising Targeted to Hispanic and Black Youth: Contributing to Health Disparities* [online]. Available at: www.uconnruddcenter.org/files/Pdfs/272-7%20%20Rudd_Targeted%20

Marketing%20Report_Release_081115%5B1%5D.pdf (accessed 22 January 2016).

Hattersley, L., Isaacs, B., and Burch, D. (2013) Supermarket Power, Own-labels, and Manufacturer Counterstrategies: International Relations of Cooperation and Competition in the Fruit Canning Industry. *Agriculture and Human Values*, 30(2), 225–233.

Irish Times. (2007) Gerard Cleary Glenisk Winning Business Worldwide. *Irish Times*. 25 June 2007, 17.

IUF. (2012) Nestlé's 'New Reality' Looks a Lot Like the Old: Squeezing Workers to Squeeze Out Cash [online]. Available at: http://cms.iuf.org/?q=node/1532 (accessed 22 January 2016).

Kellogg. (2014) *2014 Annual Report* [online]. Available at: http://investor.kelloggs.com/~/media/Files/K/Kellogg-IR/Annual%20Reports/K-2014-Annual-Report-v001-q725z5.pdf (accessed 13 April 2016).

Kerry Group. (2014) Kerry Group Corporate History [online]. Available at: www.kerrygroup.com/docs/history/Kerry_Group_Corporate_History_24-2-14.pdf (accessed 22 January 2016).

Kristof, N. (2014) The Unhealthy Meat Market [online]. *New York Times*. 12 March 2014. Available at: www.nytimes.com/2014/03/13/opinion/kristof-the-unhealthy-meat-market.html (accessed 22 January 2016).

Lawrence, F. (2004) *Not on the Label: What Really Goes into the Food on Your Plate*. London: Penguin.

McCall, B. (2013) Food Innovators have a Taste for Creating New Products. *Irish Times*. 18 February 2013, 6.

McClintock, N. (2014) Radical, Reformist, and Garden-Variety Neoliberal: Coming to Terms with Urban Agriculture's Contradictions. *Local Environment*, 19(2), 147–171.

McMichael, P. (2005) Global Development and the Corporate Food Regime. *New Directions in the Sociology of Global Development: Research in Rural Sociology and Development*, 11, 269–303.

Mondelēz. (2014) *2014 Annual Report* [online]. Available at: http://ir.mondelezinternational.com/secfiling.cfm?filingID=1193125-15-55355&CIK=1103982 (accessed 13 April 2016).

Monteiro, C.A., Gomes, F.S., and Cannon, G. (2010) Can the Food Industry Help Tackle the Growing Burden of Under-Nutrition? The Snack Attack. *American Journal of Public Health*, 100, 975–981.

Monteiro, C.A., Moubarac, J.C., Cannon, G., Ng, S.W., and Popkin, B. (2013) Ultra-Processed Products are Becoming Dominant in the Global Food System. *Obesity Reviews*, 14 (Suppl. 2), 21–28.

Moodie, R., Stuckler, D., Monteiro, C., *et al.* (2013) Profits and Pandemics: Prevention of Harmful Effects of Tobacco, Alcohol, and Ultra-Processed Food and Drink Industries. *Lancet*, 381, 670–679.

Moss, M. (2013a) The Extraordinary Science of Addictive Junk Food [online]. *New York Times*. 20 February 2013. Available at: www.nytimes.com/2013/02/24/magazine/the-extraordinary-science-of-junk-food.html?pagewanted=all&_r=0 (accessed 22 January 2016).

Moss, M. (2013b) *Salt, Sugar, Fat: How the Food Giants Hooked Us*. London: W.H. Allen.

Nestlé. (2014) *2014 Annual Report* [online]. Available at: www.nestle.com/asset-library/documents/library/documents/annual_reports/2014-annual-report-en.pdf (accessed 13 April 2016).

Nestle, M. (2013) *Food Politics: How the Food Industry Influences Nutrition and Health*. Berkeley, CA: University of California Press.

Nestle, M. (2015) *Soda Politics*. Oxford: Oxford University Press.

Neupane, S., Virtanen, P., Luukkaala, T., Siukola, A., and Nygard, C.H. (2014) A Four Year Follow-up Study of Physical Working Conditions and Perceived Mental and Physical Strain Among Food Industry Workers. *Applied Ergonomics*, 45(3), 586–591.

Newenham, P. (2012) Yogurt Manufacturer with Organic Growth. *Irish Times*. 11 September 2012, 8.

Occupational Safety and Health Administration. (2014) ConAgra Foods faces $117,000 in U.S. Department of Labor OSHA Penalties for Failing to Protect Workers from Dangerous Machinery, Other Hazards [online]. Available at: www.osha.gov/pls/oshaweb/owadisp.show_document?p_table=NEWS_RELEASES&p_id=25743 (accessed 22 January 2016).

O'Connor, A. (2014). A Call for a Low-Carb Diet That Embraces Fat [online]. *New York Times*. 1 September 2014. Available at: www.nytimes.com/2014/09/02/health/low-carb-vs-low-fat-diet.html (accessed 22 January 2016).

Office of National Statistics. (2013) *Family Spending in 2012* [online]. Available at: www.ons.gov.uk/ons/rel/family-spending/family-spending/2013-edition/info-family-spending.html (accessed 22 January 2016).

Palpacuer, F. and Tozanli, S. (2008) Changing Governance Patterns in European Food Chains: The Rise of a New Divide Between Global Players and Regional Producers. *Transnational Corporations*, 17, 69–97.

Pimentel, D. and Pimentel, M. (2008) *Food, Energy and Society* (3rd edn). Boca Raton, FL: CRC Press.

Pollan, M. (2013) *Cooked: A Natural History of Transformation*. London: Allen Lane.

Russell, N.J. and Gould, G.W. (eds) (2003) *Food Preservatives*. London: Plenum Publishers.

Sage, C. (2012) *Environment and Food*. Abingdon: Routledge.

Shand, H. (1997) *Human Nature: Agricultural Biodiversity and Farm-based Food Security* [online]. Available at: www.fao.org/sd/epdirect/epre0040.htm (accessed 22 January 2016).

Sheridan, K. (2006) An End to Farming as We Know It [online]. *Irish Times*. 26 August 2006. Available at: www.irishtimes.com/news/an-end-to-farming-as-we-know-it-1.1043609 (accessed 22 January 2016).

Strom, S. (2013) Taylor Farms, Big Food Supplier, Grapples With Frequent Recalls [online]. *New York Times*. 29 August 2013. Available at: www.nytimes.com/2013/08/30/business/taylor-farms-big-food-supplier-grapples-with-frequent-recalls.html (accessed 22 January 2016).

Tavernise, S. (2013) F.D.A. Ruling Would All but Eliminate Trans Fats [online]. *New York Times*. 7 November 2013. Available at: www.nytimes.com/2013/11/08/health/fda-trans-fats.html?pagewanted=all (accessed 22 January 2016).

Tyson Foods. (2014) A Snapshot of Tyson Foods, Inc [online]. Available at: www.tysonfoods.com/Our-Story/Tyson-Overview.aspx (accessed 22 January 2016).

UK Cabinet Office. (2008) *Food: An Analysis of the Issues* [online]. Available at: http://tinyurl.com/od6lyb6 (accessed 22 January 2016).

UNCTAD (United Nations Conference on Trade and Development). (2009) *World Investment Report: Transnational Corporations, Agricultural Production and Development*. New York, NY: United Nations.

USDA (US Department of Agriculture). (2014) Food Expenditures: Percent of Household Final Consumption Expenditures Spent on Food, Alcoholic Beverages, and Tobacco that were Consumed at Home, by Selected Countries, 2012 [online]. Available at: www.ers.usda.gov/datafiles/Food_Expenditures/Expenditures_on_food_and_alcoholic_beverages_that_were_consumed_at_home_by_selected_countries/table97_2012.xlsx (accessed 22 January 2016).

US Department of Labor. (2006) *100 Years of U.S. Consumer Spending* [online]. Available at: www.bls.gov/opub/uscs/home.htm (accessed 22 January 2016).

Villegas, P. (2013) Mexico: Junk Food Tax Is Approved [online]. *New York Times*. 31 October 2013. Available at: www.nytimes.com/2013/11/01/world/americas/mexico-junk-food-tax-is-approved.html?_r=0 (accessed 22 January 2016).

Walker, R.A. (2004) *The Conquest of Bread: 150 Years of Agribusiness in California*. London: The New Press.

Weis, T. (2007) *The Global Food Economy: The Battle for the Future of Farming*. London: Zed Books.

Young, I.M. (1990) *Justice and the Politics of Difference*. Princeton, NJ: Princeton University Press.

4 The foodscapes of selling of food

4.1 Introduction

What I tried to do in Chapter 3 was explain how the food processing industry has used organizational restructuring and food chemistry innovation to drive down the cost of production. Many of the gains here have been directed towards producing extraordinary foodstuffs, including the ultra-processed, energy-dense products that are now increasingly targeted at emerging markets in the so-called 'global south'. That third stage in the life of food is a key moment; one that generates numerous problematic, as well as some potentially more promising, features. Food processing is central to – and right in the middle of – the life of food. We do well to consider the sort of action that occurs there.

This chapter moves further downstream to the fourth stage in the life of food: to the moment when consumers spot a product and think, 'Oh, I'll have that.' All of the invention upstream of agriculture, the work on the farm or the plantation, the food engineering undertaken by the food industry, and indeed the corporate food regime overseeing all this action, has this moment in mind. On the street, in the supermarket or convenience store, the fast food outlet, or the Michelin-starred restaurant: wherever food is for sale, consumers must make a purchase, must decide to consume, for the investments in seeds, chemicals, farm machinery, labour, and new food chemistry innovations to pay off. As such, the places and spaces where food appears for sale to consumers – the foodscapes of selling food – are critical to the production and reproduction of the contemporary global food economy. For most of us, all of those earlier stages in the life of food are largely invisible and off-map relative to the up-front appearance of food around us in our daily lives. For most of us, the really apparent foodscapes are the adverts blazed in front of us – for crisps or soft drinks, for McDonald's latest invention, a salad dressing, or pasta sauce – or the micro-geographies of supermarket shelf space that seek to lure us into a sale. This chapter is about these sorts of foodscapes.

Getting to grips with the spaces where food is sold requires paying attention to many of the same themes raised in earlier chapters. There is the corporate food regime and the role of transnationals in answering the proletarian food question. There are cases of consolidation; new global food giants emerging. There is also oppression, not least because what happens here depends so intensely on the often-quite-intense exploitation of workers stacking shelves, operating check-out machines, and preparing burgers or pizzas-to-go. Moreover, as with earlier stages in the life of food, we have to consider how resistance shapes this stage in the life of food, because one crucial aspect of the contemporary food economy today is the effort on the behalf of fast-food workers to begin securing higher wages for the work they do.

Given this range of themes, the chapter is organized as follows. I begin in Section 4.2 by looking at some of the spaces and places where food is offered for sale today, focusing in particular on major developments in recent years and how these changes connect with consumers. I argue that the most significant change has been the rise of the supermarket, the 'highest temple' (Patel, 2007) of the global food economy. I therefore focus on these shops 'with everything under one roof', two- or even three-level buildings often with free parking, wifi, restaurants, and even childcare – and present them as spaces of opportunity and constraint, exhilaration and exploitation, that build and draw upon an 'architecture of territory and flow' (Pryke, 2008). I then spend some time considering fast-food restaurants as another set of important sites that shape how we understand food today. There are, though, some other sorts of spaces to consider here, especially what we might refer to as the 'alternative' foodscapes of 'foodways' and street markets, which I discuss in the case study in Section 4.3. In the final two sections I turn to consider how this stage in the life of food is oppressive, and home in on exploitation and the way some workers are resisting this: by using strikes to subvert the power of their employers and secure better working conditions.

4.2 The spaces and places of selling food

The food processing business, agriculture, the business of making and selling agri-chemicals: all such action needs food to be purchased. This is capitalism. Profit has to be made to justify all the investments, all the wages paid, all the time spent calculating what to grow, what to make, how to package, how to advertise. And if profits are going to be made, sales have to occur – people have to opt for a product; food must be bought. In this stage in the life of food, what anyone does with the food once they buy it is not so much at issue (we come to that in Chapter 5). Rather, at issue is how food is sold, hence the sorts of foodscapes we are interested in are shaped by this imperative, this need on the behalf of the whole food sector in general but then more specifically on the food sellers to makes sales happen. What we see around us, what we experience, is fundamentally shaped by this relatively odd but intense imperative.

In asking about the foodscapes in which food is bought and sold, we might immediately think of the supermarket or the restaurant and we would certainly be correct to pinpoint these as vital spaces in contemporary society. But there is one other venue we should acknowledge here: the street. Throughout the world, vast quantities of food are bought and sold on the street, whether in streets or road-side markets, or stalls near agricultural areas. Many such stalls are what we might consider to be proto-shops, i.e. they sell fruit or vegetables, maybe juices or other drinks, much like a shop does but probably without the elaborate lighting, the large fridges, or electronic payment options. Others are basically restaurants, offering freshly cooked food, sometimes for people on-the-go, but might also have covered seating areas where customers can take their time and enjoy the experience of eating. Thus, in thinking about where people buy food, these locations on the street deserve our attention: places and spaces within public space where we can buy food without having to pass money to the corporate giants who increasingly dominate the sale of food. After all, although the roasted chicken for sale at a road-side stall might well have been reared by one of the poultry sector's corporate giants, at least the owner of the stall is not KFC. Or, while much of the fruit or vegetables we might buy from a street market in an urban area might still have been produced by an agricultural firm, the final seller is not Walmart or Tesco. The street still presents us with alternatives to the humdrum, mainstream options with which so many of us are (all too) familiar today. Even so, so far as mainstream spaces are concerned, nowhere is as common as the shop or store – and today the most powerful, the one with the greatest reach, is the supermarket, as I now move on to discuss.

4.2.1 The supermarket foodscape

In the context of improved material standards of living, and the wider availability of the materials needed to formalize establishments, it is understandable that the owner of a street stall might consider doing so, perhaps by erecting walls and a roof around the food, installing lighting, placing some food in fridges, building shelves and a counter, or by beginning to use a till and bringing in a few more workers than just the owner and his or her family. This general formalizing process really took off in the nineteenth century in industrializing and urbanizing parts of the world, especially in Europe. There were the arcades that emerged in Paris in the 1840s (Benjamin, 1999); the department stores to cater for the rich in London, Chicago, or New York (e.g. see Hobsbawm, 1989: 29). The outcome was the concept of a private place where food and other (often, luxury) items were available for sale, off the street. Not only was this place away from the crowded street, it was also ordered, arranged, and made to look nice; made to feel clean, perhaps, or at least special and therefore deserving of the money spent there.

Following in the footsteps of these spaces, grocery stores soon emerged and with them came numerous new waves of innovation, which ultimately resulted in the supermarket. Evocatively referred to by Raj Patel (2007) as the 'highest temple' of the global food system, the supermarket is without question a site of incredible importance. The idea for the supermarket originates in the US in the early twentieth century. The early innovators – firms such as the Atlantic and Pacific Tea Company (A&P) or King Piggly Wiggly (see also Walker, 2004: 231–236) – saw an opportunity to save on labour and change the way consumers chose their groceries by letting them fill their own basket or trolley as they strolled around the aisles of the store, rather than standing behind a counter and asking shop assistants to go get the various goods they needed. The newness was appreciated and the supermarkets expanded numerically, geographically, in terms of the range and number of goods they sold, and even their hours of operation, to the extent that now many stores are open 24 hours per day, giving their customers ultimate 'freedom' to shop when they want.

Today, then, supermarkets are the dominant players in the global food economy. In terms of size, the largest of the supermarket chains, Walmart, dwarfs all of the other leading agriculture-based, supplier, or processing firms that we have come across so far. In 2014, for example, Walmart had global sales of $473 billion, five times those of Nestlé, the largest food processing firm (Nestlé, 2014; Walmart, 2014). Without question, their size is a result of continuous rounds of investment in new stores and warehouses, distribution networks, and logistical expertise. Like the food giants, furthermore, the supermarkets have grown via consolidation. Tesco has acquired numerous smaller firms, such as William Low, a chain of supermarkets in Scotland, which it bought in 1994 for £257 million. Then there are cases such as the Co-operative (or Co-op) Food group, which bought firms such as Local Plus, Quality Fare, Alldays, GT Smith, and Somerfield. As a result of these buyouts, the UK now has 23 large supermarket chains, which is suggestive of a relatively competitive market place, but the largest four (Tesco, Asda, Sainsbury's and Morrison's) have just over 72 per cent of the overall market; that is, they are immensely powerful and dominant (Butler, 2014). Even this level of control is nothing compared to some other places, perhaps most notably Australia, which stands out because just two groups, Coles and Woolworths, make up around 80 per cent of the food retail market (Business Monitor International, 2015). In the US, meanwhile, the one big giant is Walmart, which has '20 percent of the market share in dry grocery [and] 15 percent in fresh (areas like produce, meat, deli and bakery)' (Nixon, 2013). In turn, the size of the largest firms gives an impetus to consolidation among the smaller ones, as has happened recently with Netherlands-based Ahold agreeing in 2015 to merge with Delhaize (Business Monitor International, 2015).

Figure 4.1 displays data on five of the largest supermarket groups. Clearly, Walmart is the largest in terms of sales and number of employees, although

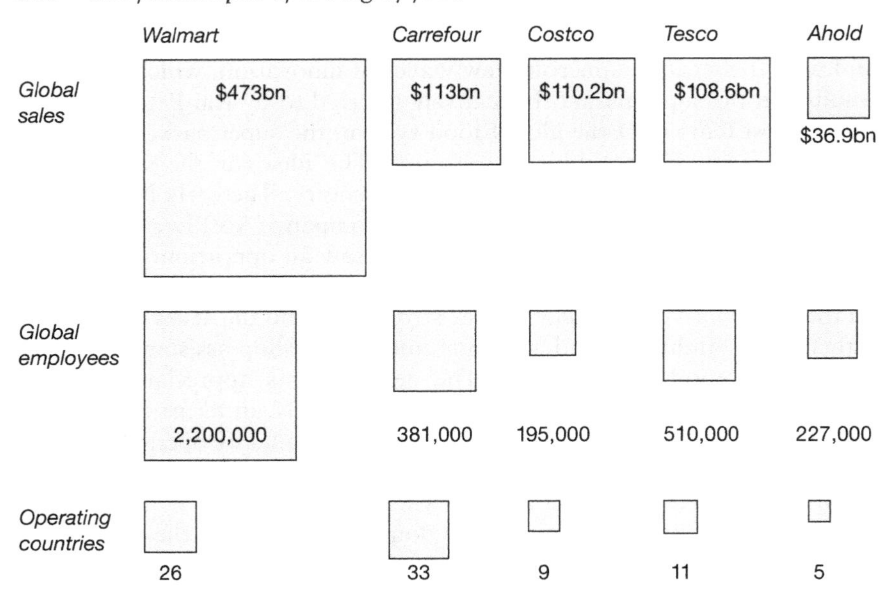

Figure 4.1 Data on selected global food retailers (sources: Ahold (2014), Carrefour (2014), Costco (2014), Tesco (2014), Walmart (2014)).

the French firm Carrefour has operations in more countries than Walmart. Walmart's international sales were $140.9 billion in 2014, demonstrating the company's expansion globally. Yet with operations in only 25 countries outside of its base in the US, further international growth seems likely. Carrefour and Tesco have similar sales, while the latter has significantly more employees. Costco has not expanded geographically as much as the other firms; yet it has much higher sales than the fifth firm, Ahold, which also has a small global footprint.

Although they are under pressure from competitors – from new entrants on the scene such as Amazon and Google, and then the rise of smaller convenience stores that are growing their market share – supermarkets are still on the rise globally. In three waves – first of all in Latin America, Central Europe, and South Africa; then in South-east Asia, Central America, and Mexico; before finally targeting China, India, and Russia – a supermarket revolution has swept across the world since the early 1990s (and not without controversy, e.g. see Franz, 2010). In each place, the share of food sold in supermarkets has soared from as low as 5 per cent to well over 50 per cent, backed up by supermarkets moving out from cities to smaller towns and rural areas, which means they are not just targeting upper classes in the city, but all workers, well-paid or not (Reardon *et al.*, 2010).

Given their size and the amount of product they can move, supermarkets have enormous buying power and this power ripples all the way through the rest of the food chain, back through the food processing industry and agriculture, even shaping the research and investment activity that happens upstream of the farm. Because supermarkets buy in bulk they can use their buying power to force suppliers to offer lower prices, which is a prominent complaint from all sorts of farmers who feel the heat from supermarkets if they cannot keep up with demand, or if they are unwilling to lower their prices. The supermarkets also demand that suppliers comply with rules and standards such as Global GAP, 'a certification and standards regime' (Pritchard, 2013: 174) developed and promoted by European supermarkets to oversee quality control in numerous aspects of agricultural production. Other suppliers, including many of the largest food processors, are also pressured by the supermarkets to drive down costs. Where supermarkets have developed their own brand labels, moreover, they exert further price pressure on food manufacturers. As such, if the proletarian food question is answered in part by what happens on the land, we have to also see the pressure supermarkets exert on the food system as part of the explanation.

This is all about their 'reach': their capacity to shape action in distant places. As the highest temple, everything that occurs in research laboratories, on farms, in food processing plants has the supermarket in mind. The point is to get product on the shelf, get it sold, get that money re-invested and reproduce the cycle, hopefully with some profits set aside for the public or private shareholders. Almost everything comes down to those few minutes when customers are inside the supermarket and choosing their items. Clearly, the supermarkets are skilled players in the capitalist economy: skilled in acquiring competitors, in borrowing effectively to pay for those acquisitions, in pursuing new markets. They now occupy such a prominent place in capitalist society, not just with regard to food. At issue for us is trying to grasp how they have come to occupy this place, as well as getting to grips with the broader significance.

4.2.2 The supermarkets' architecture of territory and flow

One way of looking at this is to suggest that the supermarkets have expanded their operations in the last few decades by creating an 'architecture of territory and flow' (Pryke, 2008). In other words, the supermarkets have developed competencies in building stores and arranging what is inside in specific ways, but also in connecting them all together. They fix some of their capital in space; creating a territory they need to protect from the intrusions of local or national regulations or from the efforts of labour unions. But their fixing of capital immediately requires mobility, a flow, a drive to get goods, people, information, and money moving through the stores. This way of imagining their actions nicely captures what supermarkets do outside and inside the actual store.

Consider what has to happen outside the store. The supermarket firms have created logistical empires that efficiently move vast quantities of goods from near and far to warehouses, stores, even to their customers' homes (see

Sage, 2012: 190–199; also De Botton, 2009). In their drive to cut costs, lower prices, and out-compete other firms, the supermarkets have had to push for quality of service, timely deliveries, order. They consequently compel actors in all other stages in the life of food to operate according to certain standards, such as the Electronic Data Interchange system, which Walmart has used extensively (Patel, 2007), and the aforementioned Global GAP (Pritchard, 2013: 174). They have scope to change practices on the other side of the world; potential to connect or drop distant suppliers into their global networks; capacity to organize and coordinate the activities of thousands of people almost instantaneously using the latest technologies and drawing on diverse knowledge about purchasing, logistics, transport, marketing. For example, with regard to fruit and vegetables, the supermarkets have created a 'permanent global summer time' (Blythman, 2004): that is, a sense for the consumer that products once only available in the summer should be available year-round, permanently – which presupposes extensive complexes in the Southern Hemisphere that operate to cater for the Northern winter (see Akram-Lodhi *et al.*, 2008).

In addition, it is worth noting here that supermarkets have developed immense infrastructures – an architecture – to identify sites for new stores, buy or rent the requisite land, secure planning permission, and ultimately build. Some such sites might be in central areas of the city, perhaps in a new shopping mall, the overall design of which they can influence by virtue of the amount of rent they will pay the property developer. In recent years, however, many supermarkets have abandoned sites inside cities and towns, preferring to build on the edge of the city, in vast retail parks in which they are the anchor tenant. Such spaces are car-oriented, not usually intended for pedestrians, and often disconnected from or poorly connected to public transport routes. The stores might have two or three floors, include restaurants, a cafe, and play facilities for children. They are locations; places to go to stock up on groceries and even to eat and relax; shops that seek to be embedded within their localities (see Coe and Lee, 2013).

In so far as many of these new giant supermarkets sell not just food but electronics, toys, clothing, hardware, and so on, they explicitly aim to out-compete other retailers in the locality. Those other stores might include another supermarket chain, but many consist of what we might refer to as 'mom-and-pop' stores; in other words, shops owned by family businesses, perhaps with relatively long histories in their towns and strong connections to the life of the place, which they might develop by sponsoring sports teams or cultural events. And so, alongside the arrival of the supermarkets and then their expansion into sales of non-food items, there has been a decline in the number of smaller stores, even to the point where supermarkets are almost the only retailer left operating.

Consequently, and for understandable reasons, many small business owners (as well as many other people) express deep concern and even outright opposition to the expansion of supermarkets. Such opposition might try to influence local decision-makers to stop planning permission or at least limit the impact of a new supermarket, perhaps by limiting its floor space. But here, too, the supermarkets are astute players, knowing very well how to work the system to their advantage, how to incorporate features that planners might appreciate, or in some cases how to buy off local officials, as Walmart has been doing in Mexico (*New York Times*, 2012). They are giants for a reason: they know their game, know how to grow, what they need to do, what to say, how to undermine their opponents. In terms of logistics, their reach, and their interest in shaping urban space, the supermarket creates a vast architecture of territory and flow: building fixed sites but ensuring those sites have what they need – that flow of goods and of customers.

Inside the store, too, these crucial makers of the capitalist economy need to create a similar architecture, one that seeks to fix and secure a space that they control and can change and re-order; but also one that manipulates the flow of goods, people, and information. Perhaps the most obvious feature to consider here is layout. Shoppers are intensely studied: how to make us spend more, how to shape what we do in the store, how to keep us there. The layout is so important because it can influence how we navigate the place. The obvious example to note here is the tendency to locate the dairy section, especially the milk, at the back of the store. But there are many other tricks. Putting the snacks near the alcohol, perhaps, or having goods in more than one place, such as a snacks aisle and at other points throughout the store, particularly at the check-out. The supermarkets have to examine how each item is sold and where it sells best, so as to re-organize the place in the best way possible, even to the extent that seasonal changes can be made (more visible fresh tropical fruit in the summer, say, or more prominently located meat for grilling). Changing layouts in this way adds to that sense of an inconstant foodscape, of nothing staying the same for too long. New aisles, new arrangements, new signage or lighting: all such changes bewilder and partially excite customers, forcing us to engage with the space, to think more carefully about why we are there, and even distracting us to the point where we unintentionally purchase items not on our shopping list.

Another technique the supermarkets have developed is forcing suppliers to compete with each other to secure the best locations. Consider here that the most expensive area in the supermarket is at the check-out where impulse purchases are made: the chewing gum we could do without, the snack bar, the magazine. But in other spaces, too, suppliers have to compete and even pay higher rates for prominence (*The Grocer*, 2013). The arrangement of the cereals or the breads on shelves might seem random, but that is far from the case, with supplier firms making premium payments to be located at eye-level or at child's eye-level. In effect, some food firms are leasing space from the supermarket, even to the point where they can determine how the goods should appear (*Modeled Behavior*, 2010).

Behind all of these efforts are other practices intended to shape our behaviour in the store. They are about shaping the store, its territory, giving it a feel, an ambience that betrays the sorts of actions that actually occur to put all of that food – all of that utterly bewildering array of food – in front of us. There is the use of music: whether to use pop music, classical music, rock, dance; whether to have it blaring out in one section but quieter in another; whether the music should be fast during the rush hour of 4–8 p.m. or slow in the morning hours. These are calculations, reflecting extensive research, that aim to shape what we do while inside the store. We are supposed to spend, not to think too much; linger around and buy on impulse. Our purchasing behaviour should be shaped, altered. If music has become one aspect in this aim, there are the other, odd, perhaps less noticeable practices. A strange one is a scattering of dirt on top of potatoes to make them look fresher, more authentic. Or, some supermarkets now have pictures of cows on the wall in meat sections, usually standing in a green pasture, probably looking young and healthy, and this despite the fact that much (if not all) of the beef for sale will have been reared in concentrated animal feed operations, never seeing a blade of grass (see Weis, 2013).

If layouts and ambience are crucial aspects of the supermarket game, a final issue is about the data we generate as supermarket customers. In the last few decades, the supermarkets have used technologies such as bar codes and loyalty cards to map customers, to understand what we want, even to predict what we might want at any one time or tempt us into buying goods the supermarkets expect we will fancy. The use of bar codes began in the 1970s in the US (Patel, 2007: 226–228). Up until then, all goods for sale in the supermarket were stamped with a sticker to indicate the price. The bar code slowly replaced stickers, which means customers today must rely on a price attached to the shelf underneath or above the item in question. The bar code might have made check-out times slightly faster, but it certainly made it easier for supermarkets to keep track of inventory: as each item was scanned and paid for, the store's database could see stock levels decline and plan for new deliveries. When matched with the loyalty card, which customers sign up to, and in doing so give away their address and other demographic data, bar codes enable a level of tracking that gives the supermarket wide-ranging surveillance capacities. It is not uncommon now to receive personalized advertisements through the mail, selling goods based on what we might have purchased before. We are tracked, intensely.

Overall, therefore, the architecture of territory and flow created by the supermarkets generates a strange, oddly emotional space familiar to us all. We feel the warm embrace of the store as we enter; the smile from a staff member waiting at the door; the signage telling us how much the store cares for us, how our health and well-being is their number one concern. In short, there's love to be found in the supermarket. We are loved for entering, loved for selecting, loved for buying. But the supermarket plays with other emotions. Knowing we might fear the grime, the dirt or the germs, the supermarket will have cleaning staff on-the-go at all times. Then there are the leaflets, signs, or info boxes on packaging, telling us just how far the store goes to look after us, to make sure we are safe. We have fears but the supermarket tries hard to allay them. Aware we are prone to deny what we know about the global food economy and how it works, the supermarkets play on this emotion, trying to help us feel that things are not so bad. The photos of cows grazing on grass, the pictures of happy staff members helping an elderly customer, the information about the locality, maybe even details of all the schools helped, and sports teams sponsored by the supermarket: these are undoubtedly intended to reinforce our own sense of denial; the sense we might develop that 'the supermarket isn't evil; it is a good place; we needn't worry'. Oddly, however, albeit with exceptions in some stores, the supermarket experience ends not with a warm goodbye but with violent rejection: rather than a casual, relaxing experience at the check-out, rather than cheers and laughter, actually leaving the store, tallying up all the goods and paying is often rapid, cold, and brutal. Our goods are swiped through the scanner – at a pace set by management, which check-out staff must keep up with – and dumped at the other end of the counter where, in a rush, we are left to bag up all the goods and get out of there before the next customer's goods start falling down from the scanner. Although retired people or school children might be there to help us bag and get out of there in supermarkets in wealthier parts of town, by-and-large the love stops when it is time to pay for the goods.

The rise of the supermarket, then, has played a major role in shaping the global food economy. It has reached the point now that people throughout the world are coming to expect that their future purchases will be made in supermarkets. They have become normalized, naturalized. Some will argue this has been a good thing for us as workers, as wage-earners, consumers. The supermarkets have compelled the food processors to think anew, to reduce costs, to manufacture food in new, cheaper ways, thereby giving us greater buying power. The supermarkets have squeezed farmers, forcing down costs, and in many cases passing on those savings to consumers (Carolan, 2012: 44–45). Certainly, all things considered, we are probably glad to have supermarkets around when we compare their prices with, say, a small convenience store. In this sense, if there is anything worse than having to spend all our money in the supermarket, perhaps it is not having a supermarket at all (e.g. on the absence of supermarkets in so-called 'food deserts', see Kurtz, 2013).

But then there is the other side of the coin to consider. The power of supermarkets has made the whole food economy so intensely focused on cost over quality or sustainability that we are at risk of rapidly losing soil, genetic diversity, knowledge about cooking, and large swathes of independent retailers that might give us an alternative to always shopping in the same place. And given that we all are their customers, we are constitutive of the same processes that ripple all the way back through the global food economy.

So long as we demand bargains – so long as we remain focused only on cost and not, say, the long-term consequences of our consumption – then we are part of the problem. Ultimately, I think it has to be said, few of us can do much about all this: not all of us have enough disposable income to opt out of the supermarket. As such, the supermarket's place in our lives is really quite cemented, locked in. The worry, though, is that our reliance on supermarkets becomes dependence and a vulnerability these large corporations can exploit. If the supermarkets have worked to perfect the art of squeezing processors and farmers (and, as we will discuss later, their workers), it seems logical to expect that squeezing their customers is firmly on the agenda.

4.2.3 The restaurant foodscape

In addition to supermarkets, the restaurant is another critical site in this fourth stage in the life of food. The idea of a restaurant is an extension of the basic principle behind the road-side food stall: someone prepares food for others who do not have the time, the skill, or inclination to cook for themselves at that particular moment. Its rise is a recent but significant development, which reflects, in part, the time many of us spend outside the home: perhaps because we commute long distances and are not always at home at meal times, or because we work away from home for a few days of the week and so need a place to eat. In part, too, restaurants cater to households in which those who cook – often, but definitely not only, women – are employed and do not always have the time to prepare food at home. Further, the expansion of restaurants must be seen as reflecting rising disposable incomes: not so much that incomes are necessarily on the rise, but once the fridge, the beds, the TV, or other consumer products have been bought, they do not need to be replaced too often and so eventually some money will become available to spend on luxury items, such as a meal out. Finally, undoubtedly, restaurants have become so significant because their owners and their suppliers have managed to drive down costs, to such an extent that, relative to eating at home, eating out is not necessarily as expensive as it might once have been.

Today, restaurants are found all over the world in various shapes and sizes. There are, in fact, 17 million restaurant outlets throughout the world and they generate annual sales of $2.3 trillion (McDonald's, 2014: 8). Many of these are no doubt family-run establishments, perhaps with just a few tables and maybe staffed only by family members. Many will be budget affairs, offering food at a reasonable price for a price-sensitive clientele. Others might be the fancy place in town, offering more of a gourmet selection, perhaps using specialist cooking techniques, higher-quality ingredients, or just offering more luxurious surroundings and a greater emphasis on service and customer experience.

Among the 17 million restaurants, one important category is known in the industry as the 'informal eating out' (IEO) segment. It includes 'quick-service eating establishments, casual dining full-service restaurants, street stalls or kiosks, cafés, 100% home delivery/takeaway providers, specialist coffee shops, self-service cafeterias and juice/smoothie bars' (McDonald's, 2014: 8). There are around eight million of these restaurants, generating annual sales in 2013 of $1.2 trillion. Big players in this segment include the global 'chain restaurant' groups such as Burger King, Pizza Hut, and McDonald's that present more or less the same menu and general spatial arrangement regardless of location. They sell so-called 'fast food': food that customers order and have on their plate within minutes. Figure 4.2 presents some data on three of the largest IEO corporations.

The fast-food restaurant stands out as a particularly striking 'innovation'. The idea that we might walk into a restaurant, order food, and have it on our tables in less than a minute or two is rather strange, given the history (albeit, a relatively short one) of restaurants in general. But when considered relative to food sold on the street, it is easy to see how it makes sense. For many people, perhaps especially in the last few decades, there simply is not enough time in the day to eat. Many employees only have a short period of time to eat, perhaps just 30 minutes. Then there are the pressures of contemporary society, the need to get a lot done in a short space of time: pay a bill, pick up light bulbs, phone a friend, and eat. Given these pressures, it is easy to understand how eating can be relegated; how it can be made into

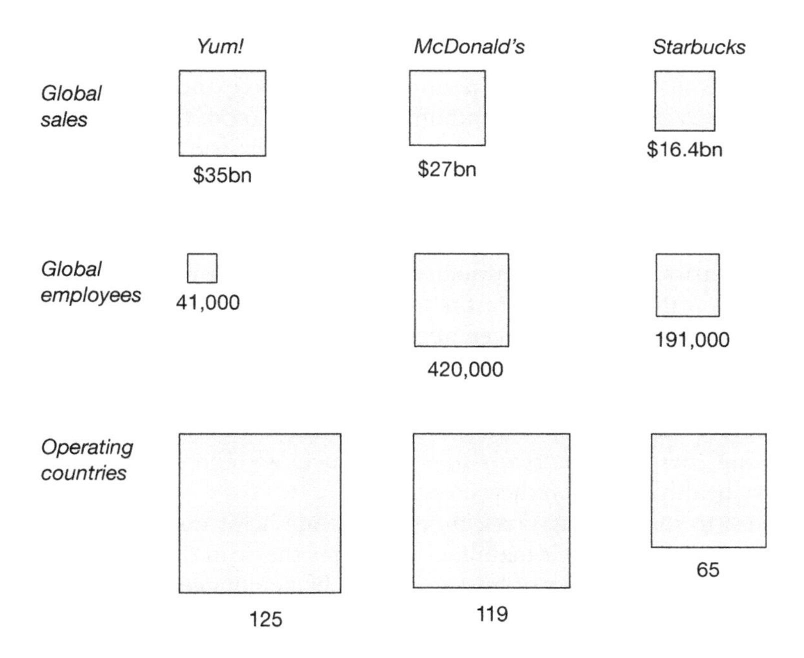

Figure 4.2 Data on selected global restaurant corporations (sources: McDonald's (2014), Starbucks (2014), Yum! (2014)).

something that does not deserve to have time set aside for it. Street food really comes in handy in such a context: the food is more or less ready; all it needs is a quick reheat, some dressing, and a plate of some sort. A hot dog, a hamburger, taco, or soup: food items that can be ordered and ready to eat in a short space of time.

Yet, over the last few decades, street food gradually has been sidelined and in its place have emerged the fast-food restaurants. They have managed to adopt and, in effect, re-imagine the idea of street food by bringing it indoors, under a roof, and with the equipment and the comfort we associate with a restaurant: a table and chairs, cutlery, warmth (or, in hot climes, cool air), perhaps someone to clean up after we have left. As we all know, the idea – not always put into practice, for one reason or another – is that we can eat quickly, using up a table but getting out of there soon (and therefore making room for someone else). There is a comfort in this. When rushed off our feet, it is hard to consider spending time to sit down and eat, to spend an hour, say, from the moment of arrival and ordering to the time we head off. The fast-food restaurant gives us this other option, but saves us the relative discomfort of eating on the street while walking, or sitting in the cold or heat, as it may be.

There is, of course, one other aspect to the fast-food restaurant's sales pitch: they are cheap. Accepting that what is cheap for one person might not be cheap for someone else, in the world of restaurants the fast-food end of the market is certainly aimed at the more budget-conscious consumer. The advertising and the special promotions tend to make this clear. A burger and fries for less than one hour's minimum wage. Chicken and fries for the same. A slice of pizza for half the minimum wage. And so on. It is fast and cheap. And it is usually low-quality and energy-dense. The fries, for example, will probably contain potato, as well as various other (and, in many cases, quite unhealthy) ingredients intended to give them a certain texture, colour, flavour, or shelf life. Likewise, the burger will certainly have some meat, but not necessarily beef, and then many other ingredients intended to keep its cost low (on this sort of ultra-processed food, see Monteiro *et al.*, 2013). The fast-food industry has been adept at imagining products and pursuing innovations that find the consumer's 'bliss point': that perfect blend of sugar, salt, and fat (see Moss, 2013). Fast food is to a large extent just an extension of the ultra-processed food industry – an industry intent on producing energy-dense, ready-to-eat products without much care for any negative health effects on their consumers.

Similar to the supermarkets, moreover, scale helps fast-food restaurants drive down costs. In their manufacturing sites they can drive down costs by producing vast quantities of burgers, fries, chicken nuggets, or pizza dough. They can buy the machinery they need at low cost by buying in bulk. Their size lets them buy energy in bulk, too, thereby helping to keep delivery costs down, at least relative to the smaller competitor. It is easy to see how the fast-food sector can out-compete the sort of 'mom-and-pop' independent restaurateur. Likewise, they are astute at arranging internal spaces in ways that draw in customers, get them out quickly, and all while minimizing costs. Most fast-food locations are cheap and cheerful. They have the easy-to-wipe plastic furniture, plastic cutlery, and bright colours keeping people awake and encouraging them to eat and leave. Recognizing that parents are budget-conscious and might be looking for a chance to relax a little, fast-food restaurants have been at the forefront of providing play spaces for children – while also aggressively marketing their products via a children's menu that includes 'free' toys or colouring books and crayons.

Thus, given their cheapness and their popularity – because, as much as they deserve criticism for their reliance on low-quality product, there is no denying the fact that many people love it – fast-food chains have major advantages over other restaurants. In turn, this advantage lets them pitch for contracts inside museums, aquariums, amusement parks, and road-side service stations. Their ubiquity then becomes almost 'path dependent'; that is, locked in, almost unavoidable, and reproduced as if it has become natural, normalized (on the 'path dependent' character of agricultural subsidies, see Carolan, 2012: 26). Via diverse tactics and strategies, fast-food restaurants have become an important feature of the global foodscapes as shaped by the corporate food regime.

4.2.4 *Alternative foodscapes*

Without question, the sorts of places and spaces I discussed above – the supermarkets, the fast-food restaurants – constitute a large part of the foodscapes of food retail. But they are certainly not the only spaces we can consider. For example, and as a major part of a broader 'alternative food movement' (see Carolan, 2012: 249–270; also Sage, 2012: 262–277), a wide range of individuals and groups are looking to give consumers a way to opt-out of buying in the supermarket. There are various types of food cooperatives offering wholesale prices to members so long as they work a few hours each month (e.g. on Brooklyn's Park Slope Food Co-op, see Ain, 2009). And in some cities community groups run bakeries or groceries that try to offer what they imagine is better-quality food than what tends to be available in the supermarket (e.g. on the People's Grocery in Oakland, see Patel, 2007: 248–252). Then there are all the many other prominent sites, such as the artisan bakery in one neighbourhood, the organic grocers in another; on one street a shop selling products imported from Italy and on another an Asian market selling vegetables, frozen fish, soy sauces, oils, herbs. Thus, in many towns and cities today there are alternative 'foodways' – alternative pathways through which consumers can buy food without having to rely on the supermarket (on Californian foodways, see Walker, 2004: 248–255; also see Guthman, 2004).

Street stalls and farmers' markets are an important sort of foodscape to consider here. Of course, as I noted earlier, in many places throughout the world people have never stopped buying and eating food on the street. Sometimes under a cover in a stall and often via even simpler arrangements – such as someone standing outside a busy office selling freshly prepared sandwiches – food has always been available for sale outside the supermarkets and restaurants. The same can also be said about urban areas in North America and Western Europe. But what does seem to be new is the growth in what are frequently referred to as farmers' markets (e.g. on the growth of farmers' markets in the US, see Brown, 2001). Indeed, in precisely the same countries where supermarkets and fast-food restaurants have become so dominant, the sight of a farmers' market, or some other form of new-wave street market, is rarely much of a surprise. We have become familiar with the idea: maybe a Thursday morning, perhaps on Saturdays – usually with some degree of support from local government, maybe also with support from some environmental groups – vendors of fruit, vegetables, or baked goods arrive and set up stalls. It is possible – and so definitely not inevitable – that many of the items available will be gourmet, high-quality, and often organic food produced locally. Much of the food will be taken home – the vegetables, say, or the meat – but lots will also be eaten there on the street, or on seats or a bench. Likewise, the vegetables, fruits, or meat for sale might be organic; they might have a better look, texture, smell, taste than what tends to be available in the supermarket. They might even be cheaper – but that is not always the point. Indeed, in many cases the products for sale will cost more than what is available in Walmart or Tesco. Rather, for many people the point is about shopping differently, sometimes in a spirit of conviviality; buying food in a way the supermarket has forgotten; interacting calmly and genuinely with the people running the stalls. It is, in part, about resisting the urge to spend in the supermarket, to avoid entering that familiar emotional place and ship a slice of your money off to their shareholders. As such, these new sorts of foodscapes call on consumers to 'check out' of the mainstream foodscape.

Understandably, these sorts of developments worry the owners of supermarkets. In response, many have introduced new products, packaged and named in 'authentic' ways, or presented in an area of the store with photos of friendly-looking farmers or small craft businesses as a backdrop. Walmart now stocks organic meat and free-range eggs; Tesco promotes their virtuous relationships with farmers. And some supermarkets go even further, such as the Whole Foods chain in the US, which has developed a market niche for offering organic food, but relying on the same architecture of territory and flow that the other, bigger supermarkets have perfected (see Zukin, 2008: 373–378). Some of the restaurant chains have also fought back. Whereas their pizzas might once have been energy-dense and generally unhealthy, many now offer a 'light' menu and some go so far as to offer organic options, all of which are presented in the new, fresh, authentic style. Even McDonald's has spotted the trend. It has changed many of its restaurants: with better-quality coffee than before, free wifi, a new 'earthier'

Figure 4.3 A downtown farmers' market, Ireland (source: the author).

design ethos replacing the bright colours of old, and the obligatory pictures of grass-fed cows.

4.3 Case study: food and gentrification

Farmers' markets are new developments that complicate our understanding of how food is sold today. In a sense, the street is back. But whether this is a cause for unequivocal celebration is another matter. In New York, for example, and as Zukin (2008: 736) notes, there is now a network of 53 open-air markets 'that provide retail outlets for small family farms' across the city. On its own, perhaps, this network seems like an exciting development, representing exactly the sort of alternative foodways and foodscape that allow consumers to avoid the supermarket. They might aim to create the sorts of spaces in which consumers can mingle, linger, and ultimately explore ways of experiencing what we might refer to as 'the new urban vocabulary' (Zukin, 2008): organic, fresh, healthy, rustic, craft, authentic. No doubt many of those who own or run the stalls will seek to create a 'feeling of community' (Zukin, 2008: 736) that should be cherished and valued. In this sense – because they offer opportunities to promote an image of a quaint, homely New York City, rather than the brutish Big Apple that TV shows and films tend to depict – street markets have become something that property journalists and tourism officials drool over.

But part of the critique of these new sorts of urban farmers' markets is about how they interact with processes of gentrification. Gentrification entails individuals and property developers investing in creating higher-value housing (e.g. see Smith, 1996). The result is often rising rents, property prices, and higher property taxes, which in turn means individuals and families on fixed or low incomes – the elderly, the working poor, the self-employed – are pushed out to other parts of the city. Thus, in so far as they aim to transform the look and feel of urban areas – and will be promoted as such – farmers' markets go some way toward attracting gentrifiers, or at least those with higher-than-average incomes. Consequently the development of these new sorts of street markets overlaps with the same exclusionary processes that make their viability possible. If the culture of gentrification is bound up with using the new urban vocabulary (Zukin, 2008) – buying organic artisan bread, Italian cheeses, or craft soy sauce – it follows that many street markets, in part because they can be relatively expensive places to buy fresh or cooked food, have become the niche spaces in which gentrifiers can spend their disposable income. Certainly, similar critiques have been made of changes in numerous other venues or sites within cities: for example, once-vibrant street markets catering to a working-class clientele, such as the Kirkgate Market in Leeds (Gonzalez and Waley, 2013), have been targeted for renewal, for renovation, by urban boosterists who see scope for replacing poorer residents of the inner city with newcomers with higher incomes, that is, exactly those who have a taste for (and can afford to buy) gourmet, craft food products (see also Low, 2011: 395–398).

In all of these sorts of ways, therefore, alternative foodscapes – such as those constituted by people who, for good reasons, might look to check out of the supermarket or avoid the chain restaurant – are just as clearly sites of struggle and sometimes as heavily contested as the mainstream foodscape. On the one hand, there is an alternative food movement, one constituted by people who might believe they can see past the visual narratives of authenticity and sustainability that are increasingly plastered all over the mainstream foodscape; one with an interest in developing solidarity with small-scale farmers, with organic producers, as well as a vibrant movement of consumers intent on trying to eat healthily. There is a new ethics of food – one caught up with notions of 'food sovereignty' (e.g. see Block *et al.*, 2012; also Wittman *et al.*, 2010) and many of the people using street markets buy into it. But on the other hand, there is an oppressive class and racial politics to the construction of these foodways (e.g. see Guthman, 2008; also Alkon *et al.*, 2013). In the North American city, for example, new street markets can combine with extant dynamics and tensions in cities where wealth is unevenly distributed and where thousands still rely on soup kitchens and government-funded food stamps (Berg, 2008). Moreover, in so far as gentrification processes can deepen patterns of racial segregation, those with higher incomes who buy food in farmers' markets might work to create alternative but also exclusionary spaces in the city. The alternative foodscape should not be glorified or celebrated uncritically. The growth of a new space economy around food has consequences, especially for those who have to move to make way for its growth.

4.4 In what sense oppressive?

Like all of the other stages in the life of food, the selling of food in mainstream or alternative foodscapes entails various forms of oppression. For example, in so far as the supermarkets seek to push to the side other venues for selling food, in so far as they build their empires to out-compete and become dominant, then we have scope to examine their practices as forms of marginalization, and even as a form of cultural imperialism. There is also a rather violent side to what supermarkets do, such as how they can abandon farmers with whom they have had long-term relationships if, say, those farmers cannot agree to price cuts. Then there is our powerlessness relative to the fast-food chains or the other chain restaurants, such as our powerlessness to limit their number, or shape what they sell, or try to sell, to us or our children. Finally, as I just noted, some of the new spaces in which food is sold, such as farmers' markets, can become bound up with violent and marginalizing processes of gentrification.

For me, though, exploitation is the particular form of oppression that really jumps out when we think about what goes into the sale of food. Recall that this refers to a transfer of the results of labour from one individual or group to another (Young, 1990). It can take place inside the home, a small, family-run workplace, or a giant TNC. It is pervasive: capitalism cannot exist without it; patriarchy draws upon it. Of course, exploitation has been lurking behind all of the earlier stages of the life of food. In research laboratories, fields, or on production lines, agricultural chemical TNCs, individual farmers or plantation owners, small bakeries or food giants such as Nestlé all necessarily rely on exploitation to produce the full range of commodities that come together – that are necessarily *brought* together, and make their way to the supermarket, the fast-food restaurant, even the street market. Yet one of the most interesting aspects about exploitation is that we tend not to recognize it as such. We certainly see labour, work; we see staff, employees, maybe rotas on the wall, maybe signs announcing the employee of the month. We might even see some practices that turn us off – maybe a manager treating an employee poorly, maybe some unsafe practices such as a staff member climbing to a top shelf without a safe stool on which to balance. Generally, however, what we encounter in the places where food is sold is people working, not so much their exploitation. But exploitation is certainly what is occurring – and on a vast scale.

Consider that in 2013 the 313 000 Tesco employees in Britain generated group profits of £2.72 billion on sales of £43.6 billion (Tesco, 2013). Tesco's management certainly played a role in making this possible; but it was the workers who ultimately moved stock to the store, put it on the shelf, kept the floors clean, and ensured customers made it out of the door. Each of those workers was paid a wage, but wages low enough that each of them on average contributed £8690 to the firm's profits. The same obviously applies to all of the profitable firms selling food. In the US, Walmart's 1.4 million workers were paid an average of $8.81, low enough that they generated $17 billion profits for the firm in 2013. These wage levels are so low that thousands of workers rely on various forms of welfare (Egan, 2014; see also Patel, 2007: 233–235).

Supermarkets have managed to become so central, so powerful, precisely because they have developed labour process techniques that make their workers more productive. At the check-out, for example, employees are monitored to see how many goods they can scan per minute or per hour. In the warehouse, teams of workers are tracked to see how many pallets of goods they can move (e.g. see Hencke, 2005). Managers, too, are given performance targets; just to make sure they exert enough pressure on their workers (e.g. see Chakrabortty, 2010). In addition to these sorts of techniques, supermarkets have mirrored practices in other sectors of the economy that are intended to squeeze productivity savings from workers. For example, numerous supermarkets have sought to ban or fight trade unions, with Walmart standing out here (Patel, 2007; also Woodman, 2012). Even where unions are present, supermarkets push to hire part-time, temporary workers and minimize the number of full-time, permanent staff members: doing so reduces what employers need to contribute towards health costs, holidays, or maternity leave. Alternatively, firms look for flexible contracts that can bring in more workers at peak times and reduce numbers when business is slow. Finally, by virtue of the low wages many such employers offer, it is often only the most vulnerable workers who take on the jobs: newly arrived immigrants, say, or women who work most of their time in the home but who need a few extra hours each week. For employers, the great advantage here is that unions tend to find it harder to organize part-time, flexible, temporary workers. If workers do not always know their colleagues too well, if they are fearful because they lack full legal status in the country or because they desperately need their pay, they might be less likely to attend a union meeting, sign a membership card, or go on strike.

Clearly, then, the sorts of places and spaces in which we buy food are produced by millions of workers, many of whom are low paid, working flexible but long hours, and lacking benefits: stacking supermarket shelves, cleaning the floors, or running the check-out. Their labour – of all the labour in the contemporary global food economy – is what many of us encounter most frequently. We might not tend to see it this way, but in many of the most common spaces where food is sold today the oppressive experience of exploitation is a definitive feature.

4.5 Resistance: fast-food workers on strike

If the dominant form of oppression in the world of food retail is exploitation, it is appropriate that we consider the strike as a key mode of resistance. Of course, the strike matters elsewhere in the life of food. Whether it is farm workers striking for better working conditions or higher pay on agricultural fields in California or Baja California in Mexico, the strike goes a long way to shaping agricultural foodscapes. Likewise, there are workers in food processing factories all over the world who strike for better conditions. But the strike really stands out in this stage in the life of food because of some recent developments in the US fast-food sector. Let me briefly discuss what this is all about.

At issue here is that wages in the fast-food sector in the US are notoriously low, often at or just above minimum wage. Indeed, in New York, where there are 55 000 fast-food workers, the minimum wage in 2013 was $7.25 an hour and the median wage in the fast-food sector was $8.90 (Bittman, 2013). Moreover, because employers compel workers to accept part-time hours – hence between one-third and one-half of New York fast-food workers need to work more than one job – they are not obliged to pay holiday or sick pay, or even cover health insurance. In New York only 13 per cent of fast-food workers receive health insurance coverage from their employers (Bittman, 2013). And then the low pay and lack of benefits means that many workers – as many as three in five in New York State – need government assistance. Indeed in the US as a whole, 44 per cent of workers in the restaurants and fast-food sector require government assistance – this is the highest rate of any industry in the US (Allegretto *et al.*, 2013). Low wages and poor working conditions, then: yet, according to a group of economists, McDonald's could cover the cost of a $10.50 minimum wage by raising the price of a Big Mac from $4 to $4.05 (Clifford, 2013).

In response, therefore, some fast-food workers have begun to fight back. With support from the Service Employees International Union, a group of workers called Fast Food Forward set out to unionize the industry and begin a call for a $15 minimum wage (Bittman, 2013). The movement began with simple protests outside a McDonald's restaurant in New York, but it has spread across the country (Clifford, 2013). The rallying call has been 'Fight for 15' (FF15) – a $15 minimum wage – and it has emboldened striking workers and supporters to continue pushing for better conditions in the fast-food sector. In Chicago, for example, the Workers Organizing Committee of Chicago, which was formed in 2012, now leads the FF15 campaign (McQuade, 2015).

However, moves to unionize the industry face numerous challenges. One key obstacle is that most fast-food restaurants are franchises. Organizing such a fragmented workforce is not easy. Moreover, the four million fast-food workers in the US are scattered across 200 000 locations (Clifford, 2013). To combat this, therefore, the Service Employees International Union has adopted what McQuade (2015) calls a 'metro strategy' which tries to mobilize workers in diverse sectors across a whole city. One-day 'flash strikes' are a key part of this strategy. The strikes raise awareness and exert pressure on employers. They have been relatively successful. McDonald's, for example, agreed to raise wages in the 1500 US restaurants it operates; however, 90 per cent of its restaurants are franchises and are therefore not covered by the wage rise (*Reuters*, 2015a). The state of New York, meanwhile, has joined other states across the US in raising minimum wages in the fast-food and other sectors, albeit relatively slowly (fast-food workers in New York City will see a minimum wage of $15 by 2018, but the minimum for workers in the rest of the state will only reach $15 by 2022) (*New York Times*, 2015).

Other successes include new links between fast-food workers and a wide range of elected representatives, as well as the general public and even groups such as UnitedNY.org and the Black Institute in New York. In Chicago, moreover, campaigners in FF15 have tried to establish connections with groups such as the Black Youth Project and Black Lives Matter, both of which demand racial justice for black people (McQuade, 2015). The point is that the fast-food sector – like other low-wage sectors – intersects with structural racism in profound ways: segregation and discrimination, for example, drive disproportionate numbers of blacks and Latinos into low-wage employment, from which many will require dependence on government support and develop exposure to various forms of vulnerability (continuing to put up with contingent work, say, not to mention hunger, disease, and shorter life expectancy) (e.g. see Block *et al.*, 2012; Alkon *et al.*, 2013; also Garcia and Sharif, 2015). As such, movements calling for higher wages can overlap, sometimes in fruitful ways, with related struggles for racial justice. Higher wages address poverty; new alliances in civil society, forged via strikes, can go some way toward undermining structural racism.

Overall, then, these workers have helped shine a light on rates of exploitation and general working conditions in some of the most frequented and everyday foodscapes. Moreover, they have helped raise awareness of another key aspect of the contemporary global food economy: the food system works well for the big shareholders of the top firms, including institutional investors such as pension funds; the rich families who still own large chunks of the stock of publicly listed firms; the families running private firms; or the top-tier managers and executives earning six- or seven-figure salaries. It works well for owners and executives such as Hugh Grant, Chief Executive Officer of Monsanto, who makes around $13 million each year (*Reuters*, 2015b); Paul Bulcke, Chief Executive Officer of Nestlé, who earned $10.6 million in 2013 (Shotter, 2014); or C. Douglas McMillon, the Chief Executive Officer of Walmart, who earned $25.6 million in 2014 (Huddlestone, 2015). These sorts of people are the real winners from the exploitation that produces the global food economy. Their high compensation rates stand in stark contrast to the workers on the ground. Capitalism is exploitative; we know this. Yet, at least in the foodscapes we find in fast-food restaurants (and other sectors of the economy where wages are equally low), the level of exploitation is now in question. In the process, as (potential) customers, we are faced with a question: to what extent might our willingness to check out of the mainstream foodscape connect with, support, or even undermine the drive to alter these rates of exploitation?

What else matters here?

I suspect my failure to spend any time dealing with the everyday foodscapes of small towns or neighbourhoods within cities might have been frustrating for you. Had I lingered about in these foodscapes, what might have leapt out as worth discussing, do you think?

I did not dwell on the idea of food deserts – that is, the idea that people living in some areas have no access at all to fresh, nutritious food. If you had to do some research on food deserts near where you live, what do you think you would find?

What if my focus had been on cultural imperialism, rather than exploitation as a form of oppression in this stage in the life of food? How might I have explored it with regard to the supermarket and fast-food restaurant scene?

You decide

- If they are the 'highest temples', as Patel (2007) claims, then shouldn't we just get on with it and embrace the supermarket? Why bother wanting to opt out?
- Do you think people in a particular place should have the right to ban supermarkets or fast-food restaurants? Should such defensive actions have a place in today's society?
- Just how should societies like ours deal with the question of low-paid work? Is it something to be tackled by government action, or is it just 'one of those things' that we can't do much about?

Suggested reading

Alain de Botton's (2010) *The Pleasures and Sorrows of Work* has a strong focus on food and how it makes its way to us, not least via the supermarket, which he creatively dwells upon in chapter 2.

Chapter 9 of Michael Carolan's (2011) *The Real Cost of Cheap Food* and chapter 8 of Raj Patel's (2007) *Stuffed and Starved* both present clear and useful discussions of supermarkets and the challenges they present for the overall food system.

In a rich and insightful edited collection of work on the 'geographies of race and food' (Slocum and Saldanha 2013), Hilda Kurtz (2013) has an excellent chapter on the connections between food deserts and racial segregation.

References

Ahold. (2014) *Annual Report* [online]. Available at: www.ahold.com/web/file?uuid=5b97bc38-30f7-4506-9894-c6c27dbb9b5f&owner=f6216a8f-4a2d-494f-8168-ae6cd1765756&contentid=2642 (accessed 22 January 2016).

Ain, A.J. (2009) Flunking Out at the Food Co-op [online]. *New York Times*. 23 October 2009. Available at: www.nytimes.com/2009/10/25/nyregion/25coop.html?_r=0 (accessed 22 January 2016).

Akram-Lodhi, A.H., Kay, C., and Borras, S.M. Jr. (2008) The Political Economy of Land and the Agrarian Question in an Era of Neoliberal Globalization. In: Akram-Lodhi, A.H. and Kay, C. (eds), *Peasants and Globalization: Political Economy, Rural Transformation and the Agrarian Question*. London: Routledge, pp. 214–238.

Alkon, A.H., Block, D., Moore, K., Gillis, C., DiNuccio, N., and Chavez, N. (2013) Foodways of the Urban Poor. *Geoforum*, 48 (August), 126–135.

Allegretto, S.A., Doussard, M., Graham-Squire, D., Jacobs, K., Thompson, D., and Thompson, J. (2013) *Fast Food, Poverty Wages: The Public Cost of Low-Wage Jobs in the Fast Food Industry* [online]. Available at: http://laborcenter.berkeley.edu/pdf/2013/fast_food_poverty_wages.pdf (accessed 22 January 2016).

Benjamin, W. (1999) *The Arcades Project*. Cambridge, MA: Harvard University Press.

Berg, J. (2008) *All You Can Eat: How Hungry is America?* New York, NY: Seven Stories.

Bittman, M. (2013) Fast Food, Low Pay [online]. *New York Times*. 25 July 2013. Available at: http://opinionator.blogs.nytimes.com/2013/07/25/fast-food-low-pay/ (accessed 22 January 2016).

Block, D.R., Chavez, N., Allen, E., and Ramirez, D. (2012) Food Sovereignty, Urban Food Access, and Food Activism: Contemplating the Connections Through Examples from Chicago. *Agriculture and Human Values*, 29(2), 203–215.

Blythman, J. (2004) *Shopped: The Shocking Power of British Supermarkets*. London: Harper Collins.

Brown, A. (2001) Counting Farmers Markets. *Geographical Review*, 91(4), 655–674.

Business Monitor International. (2015) *Food & Drink Report* [online]. 1 October 2015. Available at: www.bmiresearch.com/food-drink (accessed 22 January 2016).

Butler, S. (2015) Sainsbury's Becomes UK's Second-Biggest Supermarket. *Guardian* [online]. 28 July 2015. Available at: www.theguardian.com/business/2015/jul/28/sainsburys-becomes-uks-second-biggest-supermarket (accessed 13 April 2016).

Carolan, M. (2011) *The Real Cost of Cheap Food*. Abingdon: Earthscan.

Carolan, M. (2012) *The Sociology of Food and Agriculture*. Abingdon: Earthscan.

Carrefour. (2014) *Annual Report* [online]. Available at: www.carrefour.com/sites/default/files/RA_2014_EN.pdf (accessed 22 January 2016).

Chakrabortty, A. (2010) Why Our Jobs are Getting Worse [online]. *Guardian*. 31 August 2010. Available at: www.theguardian.com/commentisfree/2010/aug/31/why-our-jobs-getting-worse (accessed 22 January 2016).

Clifford, S. (2013) Walmart Strains to Keep Aisles Stocked Fresh [online]. *New York Times*. 3 April 2013. Available at: www.nytimes.com/2013/04/04/business/walmart-strains-to-keep-grocery-aisles-stocked.html?_r=0&pagewanted=all (accessed 22 January 2016).

Coe, N.M. and Lee, Y.S. (2013) 'We've Learnt How to be Local': The Deepening Territorial Embeddedness of Samsung-Tesco in South Korea. *Journal of Economic Geography*, 13(2), 327–356.

Costco. (2014) *Annual Report* [online]. Available at: http://phx.corporate-ir.net/External.File?item=UGFyZW50SUQ9MjY0ODcxfENoaWxkSUQ9LTF8VHlwZT0z&t=1 (accessed 22 January 2016).

De Botton, A. (2009) *The Pleasures and Sorrows of Work*. London: Hamish Hamilton.

Egan, T. (2014) The Corporate Daddy: Walmart, Starbucks, and the Fight Against Inequality [online]. *New York Times*. 19 June 2014. Available at: www.nytimes.com/2014/06/20/opinion/timothy-egan-walmart-starbucks-and-the-fight-against-inequality.html (accessed 22 January 2016).

Franz, M. (2010) The Role of Resistance in a Retail Production Network: Protests Against Supermarkets in India. *Singapore Journal of Tropical Geography*, 31(3), 317–329.

Garcia, J.J. and Sharif, M.Z. (2015) Black Lives Matter: A Commentary on Racism and Public Health. *American Journal of Public Health*, 105(8), E27–E30.

Gonzalez, S. and Waley, P. (2013) Traditional Retail Markets: The New Gentrification Frontier? *Antipode*, 45(4), 965–983.

The Grocer. (2013) Tesco Suppliers Asked to Pay for Eye-Level Display [online]. *The Grocer*. 26 October 2013. Available at: www.thegrocer.co.uk/channels/supermarkets/tesco/tesco-suppliers-asked-to-pay-for-eye-level-display/350985.article?redirCanon=1 (accessed 22 January 2016).

Guthman, J. (2004) *Agrarian Dreams: The Paradox of Organic Farming in California*. Berkeley, CA: University of California Press.

Guthman, J. (2008) Bringing Good Food to Others: Investigating the Subjects of Alternative Food Practice. *Cultural Geographies*, 15(4), 431–447.

Hencke, D. (2005) Firms Tag Workers to Improve Efficiency [online]. *Guardian*. 7 June 2005. Available at: www.theguardian.com/technology/2005/jun/07/supermarkets.workandcareers (accessed 22 January 2016).

Hobsbawm, E. (1989) *The Age of Empire*. New York, NY: Vintage.

Huddlestone, T. (2015) A Wal-Mart Worker Making $9 an Hour Would Have to Work 2.8 Million Hours to Match the CEO's Pay [online]. *Fortune*. 19 February 2015. Available at: http://fortune.com/2015/02/19/wal-mart-wage-hike-2-million-hours (accessed 22 January 2016).

Kurtz, H. (2013) Linking Food Deserts and Racial Segregation: Challenges and Limitations. In: Slocum, R. and Saldanha, A. (eds), *Geographies of Race and Food: Fields, Bodies, Markets*. Farnham: Ashgate, pp. 248–264.

Low, S. (2011) Claiming Space for an Engaged Anthropology: Spatial Inequality and Social Exclusion. *American Anthropologist*, 113(3), 389–407.

McDonald's. (2014) *Annual Report* [online]. Available at: www.aboutmcdonalds.com/content/dam/AboutMcDonalds/Investors/McDonald's%202014%20Annual%20Report.PDF (accessed 22 January 2016).

McQuade, B. (2015) A United Front: A Strong Alliance Between Fight for 15 and Black Lives Matter Would Propel Both Movements Forward [online]. *Jacobin*. 2

September 2015. Available at: www.jacobinmag.com/2015/09/fast-food-forward-black-lives-matter-police-labor (accessed 22 January 2016).

Modeled Behavior. (2010) The Economics of Supermarket Shelves [online]. *Modeled Behavior*. 6 February 2010. Available at: http://modeledbehavior.com/2010/02/06/the-economics-of-supermarket-shelves (accessed 13 April 2016).

Monteiro, C.A., Moubarac, J.C., Cannon, G., Ng, S.W., and Popkin, B. (2013) Ultra-Processed Products are Becoming Dominant in the Global Food System. *Obesity Reviews*, 14 (Suppl. 2), 21–28.

Moss, M. (2013) *Salt, Sugar, Fat: How the Food Giants Hooked Us*. London: W.H. Allen.

Nestlé. (2014) *Annual Report* [online]. Available at: www.nestle.com/asset-library/documents/library/documents/annual_reports/2014-annual-report-en.pdf (accessed 22 January 2016).

New York Times. (2012) Wal-Mart Abroad: How a Retail Giant Fueled Growth with Bribe [online]. *New York Times*. Available at: www.nytimes.com/interactive/business/walmart-bribery-abroad-series.html (accessed 22 January 2016).

New York Times. (2015) New Minimum Wages in the New Year [online]. *New York Times*. 26 December 2015. Available at: www.nytimes.com/2015/12/27/opinion/sunday/new-minimum-wages-in-the-new-year.html (accessed 22 January 2016).

Nixon, R. (2013) Billionaires Received U.S. Farm Subsidies, Report Finds [online]. *New York Times*. 7 November 2013. Available at: www.nytimes.com/2013/11/07/us/billionaires-received-us-farm-subsidies-report-finds.html (accessed 22 January 2016).

Patel, R. (2007) *Stuffed and Starved: Markets, Power and the Hidden Battle for the World Food System*. London: Portobello.

Pritchard, B. (2013) Food Chains. In: Murcott, A., Belasco, W., and Jackson, P. (eds), *The Handbook of Food Research*. London: Bloomsbury, pp. 167–176.

Pryke, M. (2008) Making Finance, Making Worlds. In: Clark, N., Massey, D.B., and Sarre, P. (eds), *Material Geographies: A World in the Making*. London: Sage, pp. 57–104.

Reardon, T., Timmer, C.P., and Minten, B. (2010) Supermarket Revolution in Asia and Emerging Development Strategies to Include Small Farmers. *Proceedings of the National Academy of Sciences*, doi: 10.1073/pnas.1003160108.

Reuters. (2015a) McDonald's Raising Average Worker Wage to About $10 an Hour [online]. *Reuters*. 2 April 2015. Available at: www.reuters.com/article/us-mcdonalds-minimumwage-idUSKBN0MS5A220150402 (accessed 16 February 2016).

Reuters. (2015b) Monsanto: People [online]. *Reuters*. Available at: www.reuters.com/finance/stocks/officerProfile?symbol=MON&officerId=162978 (accessed 22 January 2016).

Sage, C. (2012) *Environment and Food*. Abingdon: Routledge.

Shotter, J. (2014) Nestlé Cuts Chief Executive Paul Bulcke's Pay Amid Swiss Scrutiny [online]. *Financial Times*. 11 March 2014. Available at: http://on.ft.com/1ixsmO4 (accessed 22 January 2016).

Slocum, R. and Saldanha, A. (eds) (2013) *Geographies of Race and Food: Fields, Bodies, Markets*. Farnham: Ashgate.

Smith, N. (1996) *The New Urban Frontier: Gentrification and the Revanchist City*. London: Routledge.

Starbucks. (2014) *Annual Report* [online]. Available at: http://phx.corporate-ir.net/External.File?item=UGFyZW50SUQ9NTY3NTYwfENoaWxkSUQ9MjY4MTc0fFR5cGU9MQ==&t=1 (accessed 22 January 2016).

Tesco. (2013) *Annual Report* [online]. Available at: https://files.the-group.net/library/tesco/annualreport2013/pdfs/tesco_annual_report_2013.pdf (accessed 22 January 2016).

Tesco. (2014) *Annual Report* [online]. Available at: www.tescoplc.com/files/pdf/reports/ar14/download_annual_report.pdf (accessed 13 April 2016).

Walker, R.A. (2004) *The Conquest of Bread: 150 Years of Agribusiness in California*. London: The New Press.

Walmart. (2014) *Annual Report* [online]. Available at: http://stock.walmart.com/files/doc_financials/2014/Annual/2014-annual-report.pdf (accessed 22 January 2016).

Weis, T. (2013) *The Ecological Hoofprint: The Global Burden of Industrial Livestock*. London: Zed Books.

Wittman, H., Desmarais, A.A., and Wiebe, N. (eds) (2010) *Food Sovereignty: Reconnecting Food, Nature & Community*. Oxford: Pambazuka.

Woodman, S. (2012) Labor Takes Aim at Walmart – Again [online]. *The Nation*. 4 January 2012. Available at: www.thenation.com/article/165437/labor-takes-aim-walmart-again (accessed 22 January 2016).

Young, I.M. (1990) *Justice and the Politics of Difference*. Princeton, NJ: Princeton University Press.

Yum! (2014) *Annual Report* [online]. Available at: www.yum.com/annualreport/pdf/2014yumAnnReport.pdf (accessed 22 January 2016).

Zukin, S. (2008) Consuming Authenticity: From Outposts of Difference to Means of Exclusion. *Cultural Studies*, 22(5), 724–748.

5 Food consumption

5.1 Introduction

What I tried to do in Chapter 4 was depict a set of crucial yet odd foodscapes where food is sold, especially in supermarkets and fast-food restaurants, but also in the alternative foodways such as farmers' markets. I examined the way dominant food retailers create architectures of territory and flow to ensure products get to market and consumers come across products. And I gave an account of the role these actors play in driving down the price we pay for food, whether by reaching upstream to squeeze farmers or by turning the screws on workers. I think the fourth stage in the life of food is fundamentally important and that the foodscapes we encountered speak volumes about the sort of food economy we find around us (and which we play a part in creating).

In this chapter I consider the fifth and final stage in the life of food. I organize the discussion around what I call the 'open metabolic encounter', which refers to the open and emerging relationships between individuals and food that are in evidence in global foodscapes. In the first place, and as I consider in Section 5.2, this calls attention to practices, such as cooking and eating, which work to make food meaningful in diverse ways. Cooking is an everyday practice performed for numerous reasons, many of which are bound up with notions of pride, home, and identity. It is, therefore, part of the open metabolic encounter because it is an emerging practice creating multiple possibilities. Paying attention to cooking allows us to shine a light on some positive and enriching experiences that arise via our relationship with food.

My focus then turns to consider some slightly more problematic relations emerging from the open metabolic encounter. At issue is the way food gets digested; about how, in a sense, it lingers on, shapes our bodies, and then interacts with the world around us. My discussion will consider how eating necessarily creates openings for the body to interact with food. The body has to metabolize food – break it down and turn it into energy. This metabolic encounter is unpredictable because there is a deeply complex relationship between what the body needs at any one time and what the food we eat actually contains. Certainly, as we will see, there is a growing consensus that too much consumption of some types of food will generally lead to health problems, most notably type 2 diabetes. Yet, reflecting the openness of the metabolic encounter, there is no inevitability here, nor is there certainty that what we eat somehow has causal properties. This is complex terrain.

In the case study in Section 5.3 I build on the preceding discussion by focusing on the high diabetes prevalence rate in the Pacific Island countries. I argue that this feature of the global foodscape sheds light on the sorts of processes that are generating diabetes globally, which certainly involve individual decisions but need to be seen as embedded within structural features of the global food economy. In the final two sections of the chapter I consider how this stage in the life of food is oppressive. Using the idea of cultural imperialism, the final of Iris Young's 'five faces' of oppression, I discuss in Section 5.4 how particular ideas about what food is, and how we should come to eat it, interact with contemporary neoliberal mores about individual responsibility. Finally, I ask what might be a way of resisting the dominant tendencies identified here and consider the idea of boycotts as a way of fighting back against some of the problematic features of today's global foodscapes.

5.2 The 'open metabolic encounter'

One aspect of the open metabolic encounter is about how we relate to food. This is not quite as straightforward as it might appear. Of course, we relate to food in an immediate and obvious sense: we have to eat. As I have argued throughout this book, the fact of our universal need to eat is an important element for consideration when it comes to grappling with how the corporate food regime shapes agricultural systems, food processing, and the sale of food. The idea of the 'privatization of food security' (e.g. see Fraser, 2011) leaps out here. Put simply, it is increasingly the case today that, to eat, one must be able to pay for food. So the universal need to eat is one thing; the structural arrangements, processes, and practices that combine to deny so many people around the world the capability to buy food is quite another.

One relation we have with food is absolutely about this immediate question of universal need – and the geo-historical context in which that need is made realizable, or not. But food is also something with which we have numerous other relations. A way to come at some of these other relations is to develop a sense of what else food *means* for us, as humans, in all of our diversity. Cooking is an important aspect here. Billions of meals are cooked every day in household, community, and industrial settings. These are often *global* foodscapes in so far as many of the ingredients and materials brought together will have distant origins: Japanese rice cooked in Thailand; or tropical fruits or vegetables used in Western Europe. Likewise, of course, there might be the electricity generated in Tanzania by a South

African company, or the bottled water distributed through Mexican barrios by a French-owned firm. Cooking today is increasingly made possible by these global elements.

In numerous places, moreover, there are inevitably the meals cooked by people who have migrated long distances to where they now live: Puerto Ricans in the US (e.g. see Bowen and Devine, 2011); Punjabi Sikhs in Canada (e.g. see Oliffe *et al.*, 2010); or Filipinos in Hong Kong (e.g. Law, 2001). In these 'micro-scale' (or 'kitchen-scale') but still global foodscapes that are produced by migrants across the world, cooking can generate productive and beneficial new relations. Certainly, cooking for many of these migrants will be about that straightforward connection with food: the need to eat. But it can be about much more than this, too. There is pride to be won in cooking food from home, say. There is the development of a connection with other migrants; in using specific ingredients or in carrying out certain forms of preparation, perhaps with others (e.g. see Longhurst *et al.*, 2009). Preparing food – the foodstuffs brought together, sliced or diced, boiled or fried, or simply dressed – has the potential to create much more than a meal.

Among those who are not migrants, as well, cooking obviously holds out the potential to create diverse relations, from the excitement of a cake at a birthday party, to the rather more mundane but still significant exercise of heating up a ready-to-eat, store-bought meal (e.g. see Engler-Stringer, 2010). Cooking can be about passing on skills to children or partners. It can be about sharing knowledge of taste, preference, or forms of enjoyment. The practice of preparing food is something we will often do with love, pride, or one of many other feelings. It is an everyday practice. And as an everyday practice we have to imagine cooking as something that for many people is bound up with their sense of identity, reflecting diverse positionalities with respect to the social relations of which they are a part.

Cooking, then, is something we might be able to imagine as 'embodied' (e.g. see Carolan, 2011a): part of what we need to do to get through the day, but also potentially much more than that. The foodscapes of cooking do call attention to the universality of needing to eat (even if, for people with one of a range of eating disorders, consuming food is something with which they have a strained relationship); but they also illuminate broader relations we might develop with food. It is important that we imagine the preparing of food in this *open* way. Cooking is a fundamental part of the open metabolic encounter emerging out of, and certainly constitutive of, the global foodscapes I have been trying to trace in this book. The visceral properties of it all – as Longhurst *et al.* (2009: 334) suggest, the 'sensations, moods and ways of being that emerge [from the] sight, sound, touch, smell and taste' that all gets conjured up when we prepare food – might open up a variety of relations to home or to the transnational. But likewise, I think we can also gain a lot by paying attention to some of the ways that preparing food and cooking are necessarily open to influences or intrusion from the outside. Thinking about the open metabolic encounter means asking how we manage to get food onto our tables; and what the constraints emerging in doing so can mean for the human body, both in the sense of relations we develop with other bodies, but also in terms of the body's physical health. I now turn to consider issues of health in some more detail.

5.2.1 *The open metabolic encounter and health*

A remarkable feature of today's global foodscape is that, while around 800 million people are undernourished and another two billion people suffer from one or more micronutrient deficiencies, we are also seeing a marked increase in the number of people who are overweight or obese (WHO, 2014).[1] Since 1980, for example, rates of obesity have doubled worldwide. Now 500 million adults are obese and another 900 million are overweight. Explaining this development is not easy, precisely because the metabolic encounter is inherently open: we eat what we eat for diverse reasons, and often in the context of structures that constrain our options. Plus, we eat at the same time as other influences or intrusions shape our bodies; change in the size of our body is not only a matter of what we eat. There is, then, more to what we eat than our own decisions; I might be able to choose what I want to eat from the menu in front of me, but not what is on the menu in the first place.

For many experts on this topic, however, weight gain and obesity need to be considered in relation to an 'energy balance model' which posits that humans gain weight because they are eating too much and not exercising enough. Certainly, it seems as if more of us do lead sedentary lifestyles: working at a desk in front of a computer, say, or driving to work rather than walking or cycling. And then there is the fact that food consumption has tended to increase over time: now, indeed, average consumption in kilocalories is higher than what the WHO recommends in almost all countries of the world, even if under-nourishment persists (WHO, 2014). Moreover, there is the fact that a growing proportion of what we eat today is made up of 'energy-dense' foodstuffs containing too much sugar and fat (Nestle 2013: 8–11; Monteiro *et al.*, 2013). Then there are products such as sweetened drinks, many with more sugar in just one bottle than we should probably consume in a day – recall that the American Heart Association (2009) recommends that adults limit their sugar intake to between six and nine teaspoons per day, which equals about three or four teaspoons for children younger than eight years. The problem is that sugar is a toxin to the human body, because the liver struggles to process the fructose it contains (e.g. see Lustig, 2014). In effect, sugar is a carbohydrate and a fat all rolled into one. Clearly, though, and as a brief search through a supermarket or convenience store will soon reveal, there is a vast range of sugar-rich products for sale all around us. It is easy to consume much more than the amount of sugar the AHA recommends. As such, not eating carefully – consider the fast-food meals, sometimes containing 1000 calories; that pot of ice cream last night; those chips yesterday – just almost seems like it *has* to be part of the explanation for rising obesity rates. If someone is overweight or obese, and their diet contains these sorts of foodstuffs, the connection seems obvious.

Yet, looking only at individual decisions to explain the rise in obesity is a risky practice because it lumps blame for obesity on the obese person himself/herself without taking into account a wide range of other factors that might matter. Perhaps things would be different if agricultural subsidies did not prop up the overproduction of crops that make their way into junk food, not least corn in the US, or if tax breaks did not free up firms to target advertisements at children (Carolan, 2011b: 66–70). Alternatively, if governments used agricultural subsidies to lower the final price consumers paid for fruit and vegetables – and if governments increased taxes on ultra-processed foodstuffs – maybe more consumers would opt for healthier options.

The food processing industry also demands attention in this regard. Governments could, in theory (although it is not on the horizon in the context of dominant neoliberal ideas about regulation and other interventions), compel processors to reduce the production and marketing of energy-dense, ultra-processed foodstuffs (e.g. see Monteiro *et al.*, 2013). In their defence, these firms will no doubt argue that too much regulation stifles innovation; that they need to advertise to children to secure jobs for their workers; or that it is up to parents to make decisions about what their children eat. But then, when food regulators do raise new ideas, such as including clearer warnings on food labels that alert us to the potential health consequences of what we consume, the food giants are often quick to call foul (on nutritional labelling in the US, see Wilkening *et al.*, 1994; also Nestle, 2013, 2015). In the UK, for instance, a proposed compulsory system of food warning labels using traffic-light colours – green for 'go for it', amber for 'okay most of the time', and red for 'only once in a while' – faced considerable resistance from the food industry, although many of the big firms, such as Mars UK, Nestlé UK, PepsiCo UK and Premier Foods, eventually agreed to join a *voluntary* system (Campbell, 2013; Tran, 2014).

If we look beyond the food industry, too, there is scope to consider issues such as the socio-spatial structure of contemporary societies. There is, for example, a large body of research which explores how this socio-spatial structure gives rise to so-called 'obesogenic environments' (e.g. see Swinburn *et al.*, 1999; Steel, 2009: 244–246; and for a critique, see Guthman, 2013). Reflecting the energy balance model, this research asks why individuals eat energy-dense foodstuffs while also not exercising sufficiently. Part of the issue is simply the availability of energy-dense 'junk' food. This is about the ubiquity of junk food options, such as in schools, universities, and numerous other workplaces. The food giants have been especially astute in this regard by promoting the virtues and convenience of having numerous, large, and well-serviced vending machines and usually selling all of the various sorts of ultra-processed snacks with which we are all familiar. Sweet, carbonated drinks might well be sold alongside water nowadays, but this is a recent development. In canteens, too, the food giants have ensured their products are on sale at the check-out, in the fridge, and even on the main menu, which in many places is just reheated ultra-processed food shipped in that same day from a central industrial canteen, rather than cooked from scratch on the premises, as might have been the case at some stage. It is hard to avoid ultra-processed food, from the first days in school right through to higher education and work.

The notion of an 'obesogenic environment' points to other factors, too. In various neighbourhoods, for example, there are numerous features of life that might conceivably be considered in trying to understand why people are gaining weight. The extent to which children stay indoors to play video games is one thing. Then there are areas where crime levels are perceived to be too high; hence fear leads parents to discourage their kids from running around outside. Even in neighbourhoods where crime is not such a problem, if parks are not maintained – if swings are broken or dogs defecate on the grass without someone cleaning up after them – parents and kids might not fancy going outdoors. In addition, in many households there simply is not enough time for kids or parents to exercise as much as they might like to. Nor will there always be time or money for parents to prepare much food beyond a few ultra-processed options. Long hours, long commutes, after-hours work commitments: all such features of contemporary life must be seen as factors in generating rising rates of obesity.

Another prominent feature is the sprawling way our towns and cities are now organized: a type of urban development that reaches out from the old centre, going on and on, with strip mall after strip mall, fast-food restaurant after fast-food restaurant, into the suburbs and beyond. This is an urban form which prioritizes vehicle use, one product of which is that individuals tend to walk less and therefore burn fewer calories. The spaces we find in this sort of city might also be seen as 'obesogenic'. Consider all of the various shopping malls, with their 'food courts' selling 'junk food' – McDonald's next door to Burger King, with Domino's Pizza, Taco Bell, or some other firm along the way. Nowadays there might be some healthier options available in these areas, but they remain in the minority. It is true that no one needs to eat here; but the ubiquity, the normalcy of these zones, increases the likelihood that many will. Even outside the food courts, malls today are filled with unhealthy options, ranging from ice cream stores to bakeries, shops selling chocolate and then coffee shops and book stores that sell high-calorie drinks and snacks. Junk sells.

Beyond the energy balance model – beyond thinking about how individual choices mix with the food industry's practices or the 'obesogenic' socio-spatial structure of our cities – there are other considerations that might explain the rise of obesity. Guthman (2013), for example, notes that increased body fat and altered metabolism are associated with disruption in sleep patterns, which can occur when an individual works shifts, or when a parent is regularly woken at night by a child. Recent research also suggests that chronic stress releases the hormone cortisol, which encourages calorific intake of energy-dense foodstuffs (Guthman, 2013: 152; Lustig, 2014: 68–70). In other words, what we eat is not the only issue. When we eat and what our lives are like in the first place matters.

There are also some emerging ideas that the human body is more susceptible to becoming overweight due to exposure to environmental toxins, such as 'synthetic organic and inorganic chemicals [used] since 1940, in the form of pesticides, dies, perfumes, cosmetics, medicines, food additives, plastics, fire retardants, solvents, surfactants, and so forth' (Guthman, 2012: 954). Some of the diverse and complex processes used to make these chemicals will undoubtedly turn out to be harmful to human health and the environment, just as previous generations discovered negative effects associated with chemicals such as DDT (Carson, 1962). One possibility is that exposure to these chemicals affects the endocrine system, which secretes hormones and shapes how the body functions. In one experiment, mice exposed during gestation and immediately following birth to diethylstilbestrol – which was prescribed to many pregnant women in the 1950s and 1960s to prevent miscarriage – had significantly higher body weight in adulthood than control groups (Guthman, 2012: 954). This sort of result suggests that an 'epigenetic' effect is occurring – that is, DNA has not been altered but environmental exposure has changed its inherited function. In fact, some endocrine-disrupting chemicals (EDCs) might:

> provoke the division of fat cells into more fat cells or stimulate cells having no particular destination to become fat cells [which means] that they can induce fat creation regardless of caloric intake. Insofar as existing fat cells are not only very hard to get rid of but can produce even more fat cells, there is potential for large gains of adiposity from nonexceptional exposures to EDCs.
>
> (Guthman, 2012: 955)[2]

There is, then, a crucial point to take from this: the metabolic encounter is by no means a closed-off, fully understood affair. Instead, we have to consider the 'tenuous relationship of eating and body size' (Guthman, 2009: 1118). Rather than basing what we know on the energy balance model, assuming that we know how each human body deals with the food it encounters – that it is somehow 'a machine that processes calories in a predictable manner' (Guthman, 2012: 954) – we need to place the dominant energy balance model alongside other plausible explanations for obesity and, in turn, recognize that human health is shaped by diverse, often poorly understood factors and conditions. Furthermore, as highlighted by the sorts of factors that might produce something along the lines of an 'obesogenic environment', weight gain needs to be seen as a reflection of this open metabolic encounter: broader, open, societal-wide processes shape the conditions under which we access food and sit down to eat.

5.2.2 From obesity to diabetes

If our understanding of the open metabolic encounter is still emergent and therefore contested, it seems as if there is generally widespread agreement that overweight and obesity are closely connected to the development of type 2 diabetes, otherwise known as 'insulin-resistant' or 'adult-onset' diabetes. This is such a crucial feature of this fifth stage in the life of food because the disease is 'incurable and progressive' (Abbott, 2009: 399) and, if not treated properly, it can lead to limb amputations and blindness. Figures about the growth of the disease are striking. Whereas around 30 million had diabetes in 1985, the rate had reached 171 million by 2000 (Abbott, 2009), and now the International Diabetes Federation (2014: 50) estimates there are around 415 million patients, the vast majority with type 2 diabetes. About 75 per cent of those with diabetes live in low- or middle-income countries: places such as China (109.6 million cases), India (69.2 million), Brazil (14.3 million), Russia (14.3 million), and Mexico (11.5 million) (IDF, 2014: 52).

Growth of the disease is generating significant costs to public and private healthcare systems. In Mexico, the disease 'has been the primary cause of death among women and men since 2000 followed by coronary heart disease [and in 2009] was responsible for 77,699 deaths, representing 13.76 per cent of all deaths' (Barquera *et al.*, 2013). As such, screening for and then treating diabetes places an enormous strain on the government's healthcare budget, costing around $4 billion each year and set to rise given that seven out of ten adults are currently overweight (*Economist*, 2010). Then there are even poorer countries, such as India, where the number of people with diabetes has increased dramatically in recent years (Lipska, 2014). Treatments are too expensive for most poor people with diabetes; and India does not have a healthcare system capable of providing consistent chronic care. As such, complications from diabetes, such as organ failure, shorten life expectancy. A large burden of the disease falls on the public health system; but families also bear much of the cost.

Clearly, where you live when you develop diabetes has a massive influence on how you might experience the disease. In the US, for instance, the cost of treating type 1 and type 2 diabetes is around $320 billion per year (IDF, 2014: 17), which is around 7 per cent of all US healthcare costs; this cost could double over the next 15 years (Rosenthal, 2014). Most of the increase in cost is attributable to type 2 diabetes patients, of which there are an estimated 29 million in the US (Center for Disease Control and Prevention, 2014). And life with diabetes in a place such as the US is not easy. The pharmaceutical industry has developed close relations with doctors and healthcare providers to ensure its products – including expensive injectable drugs (Rosenthal, 2014) and insulin, which is produced by just three large and profitable companies, Eli Lilly, Sanofi, and Novo Nordisk (Lipska, 2016) – are widely recommended. Given this, lots of patients struggle to pay for their treatment.

All of these developments shape how we understand the global foodscape. The food system is definitely a key part of the issue. On the one hand, this is because there is evidence that prevalence rates of diabetes are closely associated with growth in consumption of sugar-rich food and drink products (Lustig, 2014: 126–127). Too much consumption of sugar-rich products makes weight gain more likely and contributes to the likelihood of developing type 2 diabetes. Yet as I have been trying to point out, such products are pervasive and profitable. If type 2 diabetes is an epidemic emerging from the open metabolic encounter, it must also be seen as a product of the particular way the corporate food regime has constructed a global foodscape in which sales revenue trumps public health.

On the other hand, the food system is caught up in all this because the food industry is free of worrying about paying for any of the costs of diseases such as type 2 diabetes. Instead, it can proceed as normal in the expectation that the individual, the state, or indeed the market will step in. And the market has spotted opportunities. There is money to be made in treating diabetes and there are numerous products available: sugar-free foodstuffs and drinks, instruments to measure blood sugar levels, and then the medicines and drugs to help patients manage the disease. Sections of the supermarket are now filled with diabetes goods, and the pharmaceutical companies – whose development and position within the capitalist economy is so similar to the agricultural chemical firms – have dedicated teams developing new product lines.

In addition, of course, there are all sorts of products and services promising to deliver weight loss, even though there is little to no evidence that any weight loss plan works (e.g. see Brownell, 2010). Wherever we look, it seems, we find adverts for weight-loss products, not to mention the actual places – the gyms, the clinics – we pass by in our daily lives. There is all the wide variety of fitness equipment (sold in the supermarket, of all places) that can be used at home; the latest gadgets and ideas about weight loss, from the fat zappers to the diet pills and drinks; and then the latest surgical developments such as tummy tucks or gastric bands. Television shows celebrate these areas of the economy; film stars boast about the latest diet, operation, or exercise technique. Losing weight for health reasons, perhaps because of the onset of type 2 diabetes, overlaps with normalizing ideas about beauty, the ideal body shape, and body size.

In short, there is a paradox: calls for weight loss, for an end to the 'obesity epidemic', but yet also an array of market actors offering 'treatments' and 'solutions'. And so what all this speaks to, in my view, is the real complexity at work in this final stage in the life of food. No easy answers. It all points to an open metabolic encounter defined by contested knowledge (about causes, solutions, policy) and clashing logics about public health and the power of the market.

5.3 Case study: food politics and economics in the Pacific Islands

One way to further grasp these clashing logics is to consider the case of the Pacific Islands countries and territories. Seven of the ten countries with the highest prevalence rates for diabetes in 2013 were Pacific Island nations: Tokelau (37.5 per cent prevalence rate); Federated States of Micronesia (35.0 per cent); Marshall Islands (34.9 per cent); Kiribati (28.8 per cent); Cook Islands (25.7 per cent); Vanuatu (24.0 per cent); and Nauru (23.3 per cent) (IDF, 2013). Because these are countries with small populations, we are not talking about a lot of people relative to the global diabetes population; but the rates are still incredibly high and on closer inspection they speak volumes about what is taking shape in this final stage in the life of food.

Murray (2001) notes that the islands were hit hard by two globalization waves since Europeans first traded and then began initiating colonies there from about the 1870s onwards. The first new arrivals were 'sandalwood traders, copra traders, planters, labour recruiters, settlers and missionaries' (Murray, 2001: 137), but colonizers soon looked for ways to begin establishing plantations – often of sugar and copra. The plantations generated vast social upheavals as local economies were integrated into systems over which they had no control, migration from outer islands to plantations left them largely depopulated, and the place of subsistence agriculture was eroded. When some of the Pacific Island countries became independent from the 1960s onwards – and reminiscent of the 'development dream' I discussed in Chapter 2 – their economic policies became more inward-oriented, including import substitution policies and other 'modernization' efforts, which were funded by aid donations and remittances from islanders who had migrated to countries such as Australia and New Zealand.

But as with so many other 'developing' countries in the 1980s and 1990s, the Pacific Islands were hit by a new wave of outside interventions in the form of neoliberal-style conditionality and structural adjustment programmes imposed by the World Bank, the Asian Development Bank, and donor countries, including Australia and Japan. These adjustments forced cutbacks in government spending, reduced interventions such as tariff reductions, and developed new policies to promote exports (Murray, 2001: 38; also Snowdon and Thow, 2013). Among the new ideas about exports was a move to encourage the Pacific Islands to develop non-traditional agricultural exports (NTAX), such as 'vanilla from Tonga, taro from Samoa, pumpkins from Vanuatu, pawpaws from the Cook Islands and kava from Fiji' (Murray, 2001: 140). These crops were grown in export monocultures that reduced biodiversity; and over-use of agricultural chemicals has had adverse effects on water sources and has led to soil degradation. A striking outcome of the move towards NTAX has been a loss of food security because land has been set aside for export-oriented production rather than for producing food for local use.

Another aspect of the neoliberal-style interventions has been about opening the Pacific Islands to new trade flows. The islands are targeted by many of the larger exporting countries in the region, not least Australia and New Zealand; but agreements are also seemingly about opening up the seas around the Pacific Island countries and territories to foreign fishing companies (Snowdon and Thow, 2013: 152). Food imports have grown, which has presented some new challenges. One is that 'traditional foods [have] fallen by the wayside [because they] are unable to compete with the glamour and flashiness of imported foods', according to Dr Temo K. Waqanivalu, a World Health Organization technical officer for nutrition and physical activity in Fiji (Parry, 2010). As Snowdon and Thow (2013: 152) note: 'Rice has replaced local healthy starchy staples, canned meats and fish have replaced local fish and seafood, and a variety of processed snack foods and beverages have replaced local juices and fruits.' Over-reliance on imported foods, especially calorie-rich and nutrient-poor products, complicates public health challenges in these places. For example, nutritional information on food labels varies enormously: whereas most imported food once came from Australia and New Zealand, much of it now comes from China, Malaysia, and the Philippines. And given that nutrition 'labels are not only inconsistent but often not in English, the common language spoken in most Pacific island countries' (Parry, 2010), it is hard for consumers to know exactly what they are eating.

Making matters worse for the Pacific Islands are imports of cheaper, fattier cuts of meat, such as 'mutton flaps' and turkey tails (e.g. see Gewertz, 2010; Snowdon and Thow, 2013). These low-quality cuts – flaps shipped in from New Zealand and turkey tails from the US – sell well in the Pacific Islands. Given their high fat content, public health officials link their sale to rising obesity rates. Fiji therefore banned flaps imports in 2000. But Tonga's initial interest in doing the same was shot down by lobbying from the New Zealand government (Gale, 2011); and Samoa, which banned turkey tail imports in 2007, was forced to remove the ban as part of its accession agreement to join the WTO. It has introduced other measures, including a sales ban and a 300 per cent import levy (Snowdon and Thow, 2013: 155).

Public health experts in the Pacific Islands argue people need more 'locally grown, nutritious and less energy-dense traditional foods' (Parry, 2010), rather than cheap meat imports or imports of other ultra-processed snacks and beverages, the movement of which is facilitated by trade agreements. Such agreements might work well for exporters in New Zealand and the US, but they limit the 'policy space' (Snowdon and Thow, 2013: 155) for governments to make democratic decisions about how to deal with public health issues. And additional pressure is now rising via the Trans-Pacific Partnership (TPP) trade agreement, which is a broader and more ambitious attempt to open export markets in the wider region (e.g. see Friel *et al.*, 2016; also *Reuters*, 2015). Food industry lobbying has looked to ensure the TPP agreement will supersede national food regulations, as well as reduce tariffs and quotas to 'facilitate integration of food supply chains' (Friel *et al.*, 2016: 10). In crafting frameworks and agreements that suit their interests, they shape the open metabolic encounter in diverse and, as public health experts might say, worrying ways.

It is, I think, also worth noting here that these sorts of developments in the Pacific Islands are not isolated. In Mexico, for example, the North American Free Trade Agreement (NAFTA), which was introduced in 1994, enabled US- and Canada-based producers to increase their exports of corn, sugar, and snack foods (e.g. see Clark *et al.*, 2012; also Carolan, 2011b: 76). In a sense, there is nothing surprising about this. Trade agreements are about opening distant places to all sorts of new flows, including products that might cause concern for public health experts. But although the logic of free trade in agricultural products and foodstuffs – a logic held together by the industry associations and individual transnationals benefiting from the corporate food regime – is about treating food like any other commodity, it seems churlish not to recognize how food is about much more than exchange, profit, and accumulation. Food has other meanings. Its place within society gives it a degree of significance that trade agreements could, but by their current ideological basis do not, account for. The open metabolic encounter is heavily shaped by the logics underpinning the corporate food regime.

5.4 In what sense oppressive?

What a strange world we live in: one big chunk of the global population overweight and obese, while another goes hungry on a regular basis. A world of extremes. As I think the earlier parts of this book demonstrate, both of these striking features of contemporary global society have to be seen as related, as products of the same general processes. On the one hand, for example, there are the food giants who play such a massive role in the making of the corporate food regime and generate an abundance of ultra-processed, energy-dense food. But on the other hand, it is often the same food giants who look to access new markets, taking advantage of structural adjustments and free trade rules that dismantled agricultural support systems in the 'third world', and tapping into food import dependence. A major feature of the life of food today is precisely that, at the point where it should be consumed – because this is what human society needs to do – far too many people cannot actually do so, just as many other people seem to have too much. There is a distribution problem, clearly, as well as a problem regarding the underlying political economy which allows the maldistribution of food in the first place. Exactly why is it acceptable that one-seventh of humanity goes hungry while another one-fifth is overweight?

There is a lot to consider here – and Iris Young's (1990) notion of cultural imperialism is a useful way of doing so. Young argued that cultural imperialism refers to situations in which dominant understandings of society render other views and practices invisible, less important. One such situation has arisen in the last few years. Consider dominant neoliberal notions about personal responsibility (e.g. see Brown, 2015). In general terms, explaining societal conditions via reference to individual responsibility leads to the conclusion that the poor choose to be poor by virtue of their bad life decisions, just as much as the wealthy deserve their wealth because they made good life decisions. As a Burger King advertising slogan might put it: 'Have it Your Way.' Your life, your choice, your fate. Underpinning this contemporary focus on the decisions and choices individuals make is a broader and peculiar vision of how society works; a still-emergent, still-conquering regime that denies existence of, or at least downplays, structural constraints. The emergence of this regime entails a process of cultural imperialism which seeks to 'responsibilize' (Brown, 2015) individuals for their decisions and actions: 'If you are poor, be more entrepreneurial; you have no one else to blame but yourself.'

With regard to this stage in the life of food, these sorts of neoliberal ideas about personal responsibility explain developments at the bodily scale in recent decades, particularly obesity, but then also the persistence of hunger, as products of individual failings rather than collective, society-wide issues. Hunger, from this perspective, is not represented as a structural problem, but rather as a question of individual failing: failing to grow enough food, failing to earn enough money, failing to have made the right choices in life. Obesity, meanwhile, is not presented as an outcome of the food giants producing ultra-processed food or trade rules demanding that countries remain open to flows of low-quality meat such as flaps or turkey tails, but rather as a result of individuals eating too much, failing to take into account the nutritional information available on the packaging, failing to exercise self-discipline. This is, then, a perspective which views thinness and fitness as signs of self-control, responsibility, and empowerment, and obesity as something that is 'beyond control' (Guthman, 2009: 1116) and a cost to taxpayers. The good citizen today eats out, consumes ultra-processed food – but stays thin. The cultural imperialism of the food economy – and capitalist society more generally, with its numerous marketing and advertising agencies, not to mention a film industry that generally celebrates a thin body form at every possible turn – generates these problematic outcomes in the final stage in the life of food, naturalizes them, makes a space for the market to address them, and passes on many of the costs to society as a whole. As such, really getting to grips with the global foodscapes charted in this book requires thinking about the cultural dimensions to political and economic decisions, calculations, and rationalities embedded within the corporate food regime.

More generally, this regime of cultural imperialism also seeks to displace alternative visions and understandings. While the corporate food giants market their products as good-quality, exciting, and cool, they also construct local or traditional foodstuffs as backward and undesirable. 'Progress' is canned processed meat. 'Development' is white bread. 'Modernity' is microwaveable. It is also a form of cultural imperialism that devalues non-conforming views and practices, from the sharing of seeds and peasant agriculture, to cooking from scratch and bonds of solidarity in the workplace. To grow food without use of agricultural chemicals, to think carefully about cooking, or to shop without going to the supermarket: such a presence in the global foodscape is to be deviant, dangerous, and unwelcome. Cultural imperialism in this fifth stage in the life of food taps into, interacts with, and builds upon all of the other forms of oppression I have tried to review. In sum, it makes for a problem-laden foodscape, as I hope this book has demonstrated.

5.5 Resistance: boycott

In trying to wrap things up for this chapter and the book as a whole, I want to consider something a little more positive. The general thrust of this book has emphasized the corporate food regime and the power and capacity of capitalist firms to shape the global foodscape. All along, though, I have tried to shine a light on some of the ways individuals – on their own or collectively – have sought to develop forms of resistance. I have tried to emphasize how the global foodscapes we see around us are contested. Things are no different in this final stage in the life of food.

There is considerable debate about food today, as this book has tried to point out. It almost seems as if food, as an issue on many of our minds, has never been hotter. Related to this is what I think we might refer to as a 'new' proletarian food question. As I have noted from Chapter 3 onwards, the food industry has developed some rather insidious practices, such as its mantra of 'when in doubt, add sugar' or its use of clever marketing tricks to target children. In developing its knowledge of food chemistry and operating at scale, the industry has played its part in lowering the final price consumers like you and I have to pay. It played its role in answering the 'classic' proletarian food question. What is new about today, though, is the emergence of calls for the food industry to use its expertise, experience, and scale to make less ultra-processed 'junk' food and replace it with more nutritious food, albeit while keeping costs low. This is, in part, the challenge set by influential authors like Michael Pollan (2008), as well as celebrity chefs such as Jamie Oliver. They also encourage their readers and viewers to look beyond processed food, to buy fresh and get in the kitchen and cook; but much of their wrath is still aimed at the food giants via calls for them to reduce their use of sugar and salt. These writers and chefs realize that individuals will still need to buy processed food, either because they do not always have the time to cook or because they cannot afford to do otherwise. The new food proletarian question posed to the food giants is: can they change their practices, not just in response to regulations in one jurisdiction, but globally?

Of course, just how this new food question is answered, or whether it is answered at all, remains to be seen. As I have noted earlier in this book, at issue are debates about corporate control over the world upstream of the farm; whether northern agricultural practices, and the concomitant drive to constantly cheapen food, will be expanded as per Amin's nightmare; whether regulation or firm-driven change will lead to more nutritious foodstuffs on our shelves and in our stomachs; and whether the workers serving us in supermarkets and restaurants will be better paid for their labour. However, also in the mix here is the question of whether individuals like you and me can do anything. Individual agency – the capacity of food producers and consumers to shape their lives – has to be at issue. There are structural constraints and a wide range of difficulties to consider here: as I have just noted in the previous section, it is unacceptable to imagine the world as constituted by responsibilized individuals operating without any relation to society as a whole. But it is also worth noting that societies can change when individuals and their broader households and communities (via political parties, trade unions, or social movements) try to construct new realities. This is about resistance, therefore, and in this final stage in the life of food I think the act of boycotting shows how such new realities might take shape.

Boycotts – including food boycotts – are a widely used form of resistance. In the US, for example, the African-American Housewives' League used boycotts under the banner 'Don't Buy Where You Can't Work' to highlight workplaces that only hired white people (Muhammad, 2011). Cesar Chavez and the United Farm Workers' Organizing Committee led boycotts of non-union lettuce and grapes in the US, specifically targeting 'the grape grower and merchant DiGorgio and the grocery chains Safeway and Lucky' (Walker, 2004: 294; see also Mitchell, 2013). Between the 1960s and 1980s, moreover, consumers throughout the world boycotted South African agricultural produce to protest against apartheid, all of which built on various black-led boycotts within the country (e.g. see Beinart, 2001: 230, 249–250). And today, to draw attention to the plight of Palestinians in the occupied territories, there is a campaign to boycott Israeli products, including food and other manufactured goods (e.g. see Beinart, 2012).

More generally, one of the most common boycotts is vegetarianism. Deciding not to eat meat is often about health concerns. But for many it is more especially about boycotting the meat industry. This can make a powerful statement. Meat is so pervasive, so common, that to decide not to eat it – and force others to reconsider their consumption patterns – can do a lot to create discussion and debate about the contemporary food system (Weis, 2013). The same can be said about decisions to focus on 'slow food' (e.g. see Hayes-Conroy and Hayes-Conroy, 2008). As the name might suggest, slow food is a dig at fast food and indeed it emerged in the mid-1980s in Italy as a reaction to the emerging fast-food culture.

Whereas many people were flocking to the first McDonald's restaurant in town, or beginning to buy their first frozen pizzas, advocates of slow food sought ways

> to preserve local cuisines and gastronomic traditions, including heirloom varieties of local grains and breeds of livestock [as a] way to fight back against both the loss of culture and the frenzy brought to us by fast foods, supermarkets, and corporate agribusiness.
>
> (Paarlberg, 2010: 153)

This movement has encouraged the development of numerous similar calls for eating local food, staying away from ultra-processed products, and ultimately deciding to take a stand in their roles as individuals and members of wider communities within the global food economy. For example, using the slogan 'think global, eat locally', Mike Small and others in Fife in Eastern Scotland launched the 'Fife Diet', which drew attention to the scope for consumers there to eat a healthy diet based only on goods produced in the local area (Cairns, 2007). In other places, similar 'eat local' schemes have flourished. Not all such efforts entail boycotts, as such, but the general thrust is to opt out, walk away from, and forget the mainstream food industry. Boycotts, in principle, if not in name.

Of course, one criticism of boycotts like these is that they are the preserve of the middle class, the rich, and elites. It might be a fair point. Affording to eat in a slow way, or using only seasonal or local or organic food can be expensive. Shopping in one of the 'ethical' supermarkets, such as Whole Foods in the US, costs more than going to Walmart. Organic is usually more expensive than 'conventional' meat. Then, for consumers wanting to eat slow food, there is the time needed to cook, which not all people can afford for one reason or another. In these senses, at least, boycotting fast food – like checking out of the supermarket – might only ever make a marginal difference to the overall scheme of things. Boycotts have their limits.

Then again, one of the most striking aspects of the last few years has been the extent to which knowledge of and interest in food has reached a much wider audience. We might not all be able to eat slow food, but we can take steps – and many of us have taken them – to change what we eat and how we eat. I put myself in this latter category. I'm not a foodie. I have no real trainspotter-like talent to know the ins and outs of different types of food. I don't exclusively eat anything in particular. But I have been affected by developments such as the slow food movement, the Fife Diet, the television shows on cooking, as well as the literature I have been trying to use in this book. So now I make my own bread, sometimes poorly, rather than buy from the store, although I still do so when necessary. I have dropped the sugar from my coffee (I used to have three cups per day, each with two teaspoons), and have mostly stopped drinking sweet, carbonated drinks, even though my Scottish heritage makes drinking Irn Bru almost a religious commitment. I also decided to stop buying pre-cooked sauces for pasta, opting instead to make my own using just canned or fresh tomatoes. I make my own small boycotts. I doubt any of this will achieve too much on its own, other than to benefit my health. But it is a stand, albeit a minor one. Like the point of actions making up slow food movements, it is about questioning where food is coming from, how it gets onto my plate, and how it might get there in the future. This book has tried to get to grips with similar questions. The acts of resistance we have come across – of all forms, no matter how minor – point to the possibility that alternatives to the corporate food regime might take shape. If they do, and if they are less oppressive, our planet and the societies shaping it stand to gain. The metabolic encounter is open because our future is open. There may be dominant tendencies; but equally, there is no inevitability when it comes to the making of global foodscapes in the future.

What else matters here?

I have focused on the act of boycotting as one form of resistance in this stage in the life of food. If my focus was on imagining alternatives, or going against normalized practices, what do you think I could have considered?

The case study I used in this chapter was about diabetes in the Pacific Islands. Do you think you could find similar issues emerging if the focus was on the place where you live?

Is the place where you live in some way 'obesogenic' and, if so, in what way?

You decide

- I have suggested consumer boycotts are a form of resistance against the way some food producers shape the global foodscape. What else do you think food consumers or activist groups might try to do when they find practices they want to oppose?
- Should trade deals take into account the relationship between agriculture, eating, and public health? If so, how?
- How *should* the 'new proletarian food question' be answered?

Suggested reading

Julie Guthman's work, not least some of her recent papers on the politics of obesity (2009, 2012) and then on environmental epigenetics (2013), will absolutely repay its study if you want to think some more about what I have referred to as the 'open metabolic encounter'.

On individual responsibility – or, more specifically, on the idea of 'responsibilization' and the condition of being 'responsibilized' – and its relation to neoliberalism, which connects with the cultural imperialism we find in this fifth stage in the life of food, see Wendy Brown's (2015) *Undoing the Demos*, especially chapter 4.

Jane Ogden (2013) reviews numerous relevant issues about eating in the contemporary period and points to additional research that can help you explore this theme in more detail.

Notes

1 The concept of obesity describes an individual whose body mass index (BMI) is greater than 30. BMI is calculated by dividing a person's weight in kilograms by the square of their body height (in centimetres). For example, if I weigh 108 kg and have a height of 185 cm, my BMI will be 31.6 and I will be classified as obese. If my BMI is between 25 and 30, however, I will simply be classified as overweight. See Guthman (2014) for a critique.

2 It is developing knowledge about issues like this that has forced the US Food and Drug Administration to place under scrutiny Triclosan, which is used in many antibacterial products such as soaps and toothpastes and which is understood to be an endocrine-disrupting chemical (Tavernise, 2013). Some campaigners call for Triclosan to be completely banned.

References

Abbott, E. (2009) *Sugar: A Bittersweet History*. London: Duckworth Overlook.

American Heart Association. (2009) *Dietary Sugars Intake and Cardiovascular Health: A Scientific Statement from the American Heart Association* [online]. Available at: http://circ.ahajournals.org/cgi/reprint/CIRCULATIONAHA.109.192627 (accessed 22 January 2016).

Barquera, S., Campos-Nonato, I., Aguilar-Salinas, C., Lopez-Ridaura, R., Arredondo, A., and Rivera-Dommarco, J. (2013) Diabetes in Mexico: Cost and Management of Diabetes and Its Complications and Challenges for Health Policy. *Globalization and Health*, 9(3), DOI: 10.1186/1744-8603-9-3.

Beinart, P. (2012) To Save Israel, Boycott the Settlements [online]. *New York Times*. 18 March 2012. Available at: www.nytimes.com/2012/03/19/opinion/to-save-israel-boycott-the-settlements.html (accessed 22 January 2016).

Beinart, W. (2001) *Twentieth-Century South Africa*. Oxford: Oxford University Press.

Bowen, R.L. and Devine, C.M. (2011) 'Watching a Person Who Knows How to Cook, You'll Learn a Lot': Linked Lives, Cultural Transmission, and the Food Choices of Puerto Rican Girls. *Appetite*, 56(2), 290–298.

Brown, W. (2015) *Undoing the Demos*. Brooklyn, NY: Zone.

Brownell, K.D. (2010) The Humbling Experience of Treating Obesity: Should We Persist or Desist? *Behaviour Research and Therapy*, 48, 717–719.

Cairns, G. (2007) A Cause to Diet For [online]. *Guardian*. 21 November 2007. Available at: www.theguardian.com/environment/2007/nov/21/guardiansociety supplement.ethicalliving (accessed 22 January 2016).

Campbell, D. (2013) Food Packaging 'Traffic Lights' to Signal Healthy Choices on Salt, Fat and Sugar [online]. *Guardian*. 19 June 2013. Available at: www.theguardian.com/society/2013/jun/19/traffic-light-health-labels-food (accessed 22 January 2016).

Carolan, M. (2011a) *Embodied Food Politics*. Farnham: Ashgate.

Carolan, M. (2011b) *The Real Cost of Cheap Food*. Abingdon: Earthscan.

Carson, R. (1962) *Silent Spring*. New York, NY: Houghton Mifflin.

Center for Disease Control and Prevention. (2014) *National Diabetes Statistics Report, 2014* [online]. Available at: www.cdc.gov/diabetes/pubs/statsreport14/national-diabetes-report-web.pdf (accessed 22 January 2016).

Clark, S.E., Hawkes, C., Murphy, S.M.E., Hansen-Kuhn, K.A., and Wallinga, D. (2012) Exporting Obesity: US Farm and Trade Policy and the Transformation of the Mexican Consumer Food Environment. *International Journal of Occupational and Environmental Health*, 18(1), 53–64.

Economist, The. (2010) One Taco Too Many [online]. *The Economist*. 21 October 2010. Available at: www.economist.com/node/17314636 (accessed 13 April 2016).

Engler-Stringer, R. (2010) The Domestic Foodscapes of Young Low-Income Women in Montreal: Cooking Practices in the Context of an Increasingly Processed Food Supply. *Health Education & Behavior*, 37(2), 211–226.

Fraser, A. (2011) Nothing but its Eradication? Ireland's Hunger Task Force and the Production of Hunger. *Human Geography*, 4(3), 48–60.

Friel, S., Ponnamperuma, S., Schram, A., Gleeson, D., Kay, A., Thow, A.M., and Labonte, R. (2016) Shaping the Discourse: What Has the Food Industry been Lobbying for in the Trans Pacific Partnership Trade Agreement and What are the Implications for Dietary Health? *Critical Public Health*, doi: 10.1080/09581596.2016.1139689.

Gale, J. (2011) Choosing Between Free Trade and Public Health [online]. *Bloomberg Business Week*. 22 November 2011. Available at: www.bloomberg.com/bw/magazine/choosing-between-free-trade-and-public-health-11232011.html (accessed 21 February 2016).

Gewertz, D.B. (2010) *Cheap Meat: Flap Food Nations in the Pacific Islands*. Berkeley, CA: University of California Press.

Guthman, J. (2009) Teaching the Politics of Obesity: Insights into Neoliberal Embodiment and Contemporary Biopolitics. *Antipode*, 41(5), 1110–1133.

Guthman, J. (2012) Opening Up the Black Box of the Body in Geographical Obesity Research: Toward a Critical Political Ecology of Fat. *Annals of the Association of American Geographers*, 102(5), 951–957.

Guthman, J. (2013) Too Much Food and Too Little Sidewalk? Problematizing the Obesogenic Environment Thesis. *Environment and Planning A*, 45, 142–158.

Guthman, J. (2014) Doing Justice to Bodies? Reflections on Food Justice, Race, and Biology. *Antipode*, 46(5), 1153–1171.

Hayes-Conroy, A. and Hayes-Conroy, J. (2008) Taking Back Taste: Feminism, Food and Visceral Politics. *Gender Place and Culture*, 15(5), 461–473.

IDF (International Diabetes Federation). (2013) *IDF Diabetes Atlas* (6th edn) [online]. Available at: www.idf.org/sites/default/files/EN_6E_Atlas_Full_0.pdf (accessed 22 January 2016).

IDF (International Diabetes Federation). (2014) *IDF Diabetes Atlas* (7th edn) [online]. Available at: www.diabetesatlas.org/component/attachments/?task=download&id=116 (accessed 22 January 2016).

Law, L. (2001) Home Cooking: Filipino Women and Geographies of the Senses in Hong Kong. *Ecumene*, 8(3), 264–283.

Lipska, K. (2014) The Global Diabetes Epidemic [online]. *New York Times*. 25 April 2014. Available at: www.nytimes.com/2014/04/26/opinion/sunday/the-global-diabetes-epidemic.html?_r=0 (accessed 22 January 2016).

Lipska, K. (2016) Break Up the Insulin Racket [online]. *New York Times*. 20 February 2016. Available at: www.nytimes.com/2016/02/21/opinion/sunday/break-up-the-insulin-racket.html?_r=0 (accessed 20 February 2016).

Longhurst, R., Johnston, L., and Ho, E. (2009) A Visceral Approach: Cooking 'At Home' with Migrant Women in Hamilton, New Zealand. *Transactions of the Institute of British Geographers*, 3 (July), 333–345.

Lustig, R. (2014) *Fat Chance: The Hidden Truth about Sugar, Obesity and Disease*. London: Fourth Estate.

Mitchell, D. (2013) 'The Issue is Basically One of Race': Braceros, the Labor Process, and the Making of the Agro-Industrial Landscape of Mid-Twentieth-Century California. In: Slocum, R. and Saldanha, A. (eds), *Geographies of Race and Food: Fields, Bodies, Markets*. Farnham: Ashgate, pp. 79–96.

Monteiro, C.A., Moubarac, J.C., Cannon, G., Ng, S.W., and Popkin, B. (2013) Ultra-Processed Products are Becoming Dominant in the Global Food System. *Obesity Reviews*, 14 (Suppl. 2), 21–28.

Muhammad, E. (2011) *The Real Housewives of Detroit?* [online]. 4 May 2011. Available at: www.finalcall.com/artman/publish/Perspectives_1/article_7800.shtml (accessed 22 January 2016).

Murray, W.E. (2001) The Second Wave of Globalisation and Agrarian Change in the Pacific Islands. *Journal of Rural Studies*, 17(2), 135–148.

Nestle, M. (2013) *Food Politics: How the Food Industry Influences Nutrition and Health*. Berkeley, CA: University of California Press.

Nestle, M. (2015) *Soda Politics*. Oxford: Oxford University Press.

Ogden, J. (2013) Eating Disorders and Obesity: Symptoms of a Modern World. In: Murcott, A., Belasco, W., and Jackson, P. (eds), *The Handbook of Food Research*. London: Bloomsbury, pp. 455–470.

Oliffe, J.L., Grewal, S., Bottorff, J.L., *et al.* (2010) Masculinities, Diet and Senior Punjabi Sikh Immigrant Men: Food for Western Thought? *Sociology of Health & Fitness*, 32(5), 761–776.

Paarlberg, R. (2010) *Food Politics: What Everyone Needs to Know*. Oxford: Oxford University Press.

Parry, J. (2010) Pacific Islanders Pay Heavy Price for Abandoning Traditional Diet [online]. *Bulletin of the World Health Organization*. Available at: www.who.int/bulletin/volumes/88/7/10–010710/en (accessed 22 January 2016).

Pollan, N. (2008) *In Defense of Food: An Eater's Manifesto*. New York, NY: Penguin.

Reuters. (2015) Hundreds of Thousands Protest in Berlin Against EU–U.S. Trade Deal [online]. *Reuters*. 10 October 2015. Available at: http://in.reuters.com/article/trade-germany-ttip-protests-idINKCN0S40LL20151010 (accessed 22 January 2016).

Rosenthal, E. (2014) Even Small Medical Advances Can Mean Big Jumps in Bills [online]. *New York Times*. 5 April 2014. Available at: www.nytimes.com/2014/04/06/health/even-small-medical-advances-can-mean-big-jumps-in-bills.html (accessed 22 January 2016).

Snowdon, W. and Thow, A.M. (2013) Trade Policy and Obesity Prevention: Challenges and Innovation in the Pacific Islands. *Obesity Reviews*, 14 (Suppl. 2), 150–158.

Steel, C. (2009) *Hungry City: How Food Shapes our Lives*. London: Vintage Books.

Swinburn, B., Egger, G., and Raza, F. (1999) Dissecting Obesogenic Environments: The Development and Application of a Framework for Identifying and Prioritizing Environmental Interventions for Obesity. *Preventative Medicine*, 29, 563–570.

Tavernise, S. (2013) F.D.A. Questions Safety of Antibacterial Soaps [online]. *New York Times*. 16 December 2013. Available at: www.nytimes.com/2013/12/17/health/fda-to-require-proof-that-antibacterial-soaps-are-safe.html (accessed 22 January 2016).

Tran, M. (2014) Cadbury Makes Anti-obesity Pledge with Cap on Chocolate Bar Calories [online]. *Guardian*. 3 June 2014. Available at: www.theguardian.com/uk-news/2014/jun/03/cadbury-anti-obesity-pledge-cap-chocolate-calories-mondelez (accessed 22 January 2016).

Walker, R.A. (2004) *The Conquest of Bread: 150 Years of Agribusiness in California*. London: The New Press.

Weis, T. (2013) *The Ecological Hoofprint: The Global Burden of Industrial Livestock*. London: Zed Books.

WHO (World Health Organization). (2014) *Mean Body Mass Index (BMI): Situation and Trends* [online]. Geneva: World Health Organization. Available at: www.who.int/gho/ncd/risk_factors/bmi_text/en (accessed 22 January 2016).

Wilkening, V., Dexter, P., and Lewis, C. (1994) Labelling Foods to Improve Nutrition in the United States. In: World Health Organization (ed.), *Food, Nutrition and Agriculture*. Rome: WHO.

Young, I.M. (1990) *Justice and the Politics of Difference*. Princeton, NJ: Princeton University Press.

Index